malt

A Practical Guide from Field to Brewhouse

T0164075

John Mallett

brewers publications

Brewers Publications
A Division of the Brewers Association
PO Box 1679, Boulder, Colorado 80306-1679
www.BrewersAssociation.org
www.BrewersPublications.com

Carapils®/Carafoam®, Carahell®, Carafa®, Caramünch®/Caramunich®, Carared®, Caraamber®, Caraaroma®, Carawheat®, and SINAMAR® are registered trademarks of Weyermann® Malting Company.

Printed in the United States of America.

10 9 8 7 6 5 4 3 2

ISBN-13: 978-1-938469-12-1
ISBN-10: 1-938469-12-7

Library of Congress Cataloging-in-Publication Data:

Mallett, John, 1964-
  Malt : a practical guide from field to brewhouse / by John Mallett.
     pages cm
  Includes bibliographical references and index.
   ISBN 978-1-938469-12-1 -- ISBN 1-938469-12-7 1. Malt. 2. Malt liquors. I. Title.
   TP570.M26 2014
   663'.6--dc23
                            2014038768

Publisher: Kristi Switzer
Technical Editors: Joe Hertrich, John J. Palmer
Copyediting: Oliver Gray
Indexing: Doreen McLaughlin
Production and Design Management: Stephanie Johnson Martin
Cover and Interior Design: Kerry Fannon and Justin Petersen
Cover Illustration: Alicia Buelow

To my family: Linus, Kat and Maggie.
This work is evidence of your unwavering support and patience; thanks
for making it possible.

Table of Contents

Kilning
Off-Flavors
Operation
Cleaning
The Result
Malting Losses
Moisture Management

Flavor Development
    Advanced Malt Flavor Chemistry
High-Dried Malts
Caramel Malts
Roasted Malts
Making Specialty Malts
Other Grains
Other Processes
Other Products
    Dehusked/Debittered Malts
    Roasted Unmalted Grains
    Pre-Gelatinized Adjuncts
Malt Extracts
Lagnappe

Introduction to Enzymes and Modification
Carbohydrates
Sugars
Starches
Proteins
Lipids
Browning Reactions in the Kiln and Kernel
Diastatic Power in Malts
Enzyme Action

Acknowledgments

Writing a book was a stretch for me. I am an inefficient writer and easily distracted. Quite simply, this book would not have been possible without the tremendous patience of my publisher, Kristi Switzer. Her constant encouragement and gentle prodding have helped to fundamentally shape this work. Over the past two years, as this book slowly came together, she was an indispensable asset, guiding me through the pitfalls of writing. Too often, the competing pressures of work and family pushed writing to the back of the line. She gently kept me on track, encouraging and supporting me, finding resources when they were most needed.

It was she who suggested John Palmer to help with the content of the book. In addition to his thorough technical edit, his assistance clarifying certain sections has been more than generous. Joe Hertrich, *Malt's* other technical editor, has been a source of information about malt for many, many years. His willingness and ability to share knowledge is greatly appreciated. Thanks to Oliver Gray for editing my writing, and Iain Cox who provided direction at a crucial time.

The Bell's family—Larry, Laura, and the entire staff—have been incredibly supportive over the many years I have had the pleasure of working with them. That support has provided me the chance to dig deeply into barley and malt, and made this book possible. I'd also like to offer a special thanks to the Bell's malt team: Ed Ruble, Andy Farrell, Andrew Koehring, and Rik Dellinger. You guys are awesome.

Andrea Stanley of Valley Malt pointed the way to some fantastic resources. She is a fellow malt history nerd, and with her husband Christian, is helping to reshape the small-scale malting landscape in the US. Their enthusiasm for malt is absolutely infectious, and it is a pleasure to know and work with them.

Researching this book has been an absolute dream. I had the chance to interact with plenty of smart and passionate people who gave valuable insight over the years. They include (in no particular order): Mike Turnwald, Dave Thomas, Chris Swersey, Matt Brynildson, Jennifer Talley, Wayne Wambles, Jonathan Cutler, Tom Nielsen, Pat Hayes, Paul Schwarz, Bruno Vachon, Dave Kuske, John Harris, Peter Simpson, Susan Welch, Mary-Jane Maurice, Bill Wamby, Alec Mull, Joe Short, Clay Karz, Alex Smith, Mike Davis, Scott Heisel, Sean Paxton, Yvan de Baets, Dan Carey, Gordon Strong, Scott Dorsch, Bret Manning Van Havig, Jace Marti, and Eric Toft.

By my side throughout this project has been another set of inspiring people who I have never actually met. They are the authors that have written so extensively and astutely about malt, chief among them Dennis Briggs (*Malts and Malting*) and H. Stopes (*Malt and Malting*).

Thanks to everyone for the opportunity and the journey; it's been a blast.

-John Mallett

Foreword

For decades, John Mallett and I have crisscrossed and intersected each other's careers like the warp and weft of burlap malt sacks. I first met John (virtually) when I was working in Chungcheongbuk-do, South Korea, commissioning the new Jinro-Coors tower malthouse in 1993 (where, at times, I marveled at 100 pound men carrying 100 pound sacks of local barley up the steps at the loading dock). During the "spare" time that came with the hurry-up-and-wait activities of new plant commissioning, I co-authored (with Professor Sir Geoffrey Palmer) a few articles about malt for the March-April 1994 *The New Brewer* magazine. John was technical editor for *The New Brewer* then, so my articles ended up in his inbox.

After returning to the US, I would periodically bump into John at various brewing functions. He was assisting several different breweries in Colorado and elsewhere, while I continued working for Coors in assorted R&D, malting, and brewing posts around the world, until I retired in 2007. John scaled up in brewery size and position, and is now Production Director at Bell's Brewery in Kalamazoo, Michigan. Bell's is a large, regional craft brewery that grows its own barley for malting. He also teaches at Siebel Institute of Technology and writes technical articles on all aspects of brewing.

Recently, our paths crossed again when we realized we were both writing books about malt. Mine, *The Craft Maltsters' Handbook*, recently published by White Mule Press (Hayward, California), and John's book published by the Brewers Association (Boulder, Colorado). When we bumped into each other at the 2014 Craft Brewers Conference in Denver, I asked John about possible redundancies between our two projects. He heartily replied, "don't worry; yours is written from the maltster's point of view and mine is the brewer's perspective. They will complement each other!" He was right. They do nicely.

John talks about the "heavy-lifting" that malt does for brewers. In this book, John has done the heavy lifting for us by presenting (in a very readable fashion) the chemistry of malt carbohydrates, sugars, amino acids, proteins, and lipids. Throughout the book, he elegantly describes the history and chemistry of Maillard reaction products, derivation of caramel colors and flavors in the kiln and kettle. John pulls from his own experience as a brewer, and brings in several other notaries from the craft brewing and production malting worlds to emphasize the important and sometimes surprising practical aspects of using malt to make beer. The book flows like the air moving through drying barley, describing functionality, flavors, fermentability and unfermentables extracted from malt, including how many malty factors can be unintentionally over-represented in beer. As John discusses brewing recipe formulation, one brewer he interviewed compares the process to painting. Color, depth, and brush strokes can be just like the different qualities and quantities of malts. Other brewers think of their brewing formulae as musical compositions, with different malts providing bass, middle, and treble notes.

Doing research for the book, John went on many enviable visits to malthouses and breweries around the world and neatly tours us through them. In reviewing the history of malting, John tells us about Harry Harlan, the "Indiana Jones" of barley. He tells us about the ongoing search for the next "Maris Otter" of malting barley varieties. The different styles of standard and specialty malts are introduced, and most importantly, critiqued from several brewers' perspectives. He shares his and others' "teachable moments of near catastrophe" in handling and brewing with different malts over the years. In addition to the malt itself, he addresses common concerns related to malt receiving, conveying, storing, weighing, and milling. You won't find that level of fundamental understanding and practicality in any malting theory textbooks.

In the late 1970s, when the first craft brewers were turning their homebrewing hobby into commercial businesses, Bill Coors called several of us into his office. He said that we would probably be getting technical requests from small, startup breweries and when we did, we were to "get on an airplane and go." So we did. Many of us drove, flew, or answered questions by phone whenever asked. One of the first instances I remember was from a fledgling Colorado brewer that appeared wide-eyed in my office carrying a case box full of milled malt, asking why they couldn't get good extract and run-offs from their malt. This one was dead easy. I scooped up a handful from the box and showed him the whole malt kernels, or "old maids," that should not have been there after milling. I told him how to adjust his mill, gave him target sieve percentages to shoot for, and sent him on his way. Those brewers learned quickly and have continued to study the art, brewing in Colorado now for thirty-five years. It is in the same spirit of generously sharing wisdom that John wrote this book. Brewers helping brewers. We can all learn from each other, even if we're competing against each other on the shelves!

In learning about malthouse hygiene as a Siebel student, John mentions the "white bread test" introduced to him at the Schreier (now Cargill) Malthouse in Sheboygan, Wisconsin. Mick Stewart, South African Breweries (SAB) Chief Brewer, invented this test many years ago when he was inspecting malthouses. The test simply said that any head maltster should be confident enough in their plant's hygiene that they would willingly take a bite from a slice of white bread after swiping it across any malthouse surface—inside or outside of pipes, tanks, and walls. After I showed Mick around my facility in the early 1980s, he remarked that we operated one of the few malthouses he had ever seen that could actually pass his test. This level of hygiene was achieved rather expensively, by a five-person malthouse crew who did nothing but clean. In the book, John discusses simple and cost-effective ways to keep malt handling and storage areas clean and safe in any size brewery (incidentally, Mick also told SAB packaging managers they ought to keep their pasteurizer water clean enough that they would voluntarily bathe their own babies in it!).

John rightly offers that "barley isn't as sexy as hops." Ask any maltster and she will tell you that hops seem sexier because they are easier to get into the glass of beer. Isomerization of kettle hops and flavors infused from late and dry hopping are simple, straightforward physical and chemical processes. No confounding (and confounded) biology or biochemistry (except for the anti-microbial effect of hops in beer). This rectilinear "spice in, flavor out" cooking model makes hops easier to relate to and understand for brewers and consumers alike. Everyone cooks; few malt.

Malting barley, on the other hand, must pass through additional elaborate technical processing steps including ensuring practically perfect biological viability of the cereal seed before, during, and after harvest; nurturing vigorous, uniform, and hygienic biological growth during steeping and germination; developing biscuity, nutty, toffee, vanilla, caramel, coffee, roasted, toasted, and malty aromas and flavors in the kiln; and empowering brewing horsepower in the form of one hundred-plus discrete enzymes that determine, inter alia, the fermentability, alcohol, color, mouthfeel, flavor and foam stability, productivity, throughput, and economics of the final product. All these factors may seem sexy to a maltster, but try explaining them to a punter over a pint in the pub. All of this complexity means that malts and malting are too far removed from beer for nearly all consumers and most brewers (present readers excluded) to invest time, energy, and interest in.

Woe, however, to the unwitting and uninformed brewer who chooses to circumvent the malting process by replacing malt in their brewhouse with unmalted barley and fungal or bacterial enzymes. Many moons ago I studied malting and brewing at Heriot-Watt University in Edinburgh. I would occasionally enjoy a beer or two at a local pub with a friend who brewed at a large UK brewery. He proudly proclaimed that his brewery had successfully replaced a large portion of their malt in the brewhouse (40 percent, if memory serves) with raw barley and fungal enzymes to save production costs. I told him that if I closed my eyes and tasted his beers alongside others I could always pick out his beers because of a slight raw grain off-note in the beer. I proceeded to perform this feat for him several times successfully. Within a few years, his brewery went out of business. True story; names withheld for personal protection.

Partially replacing malted grains with unmalted adjuncts in brewing is a hot-button practice that is still debated today. Is it an effort to improve beer drinkability, profitability, or both? The Brewers Association has recently (2014) changed their view on the subject by expanding the definition of craft beer to include "... *beers which use adjuncts to enhance rather than lighten flavor.*" The pros and cons of adjuncts are not within the scope of this book, and have been discussed by brewers and laypersons for decades, as this naïve diatribe published in the *Denver Daily Tribune* on October 2, 1878 typifies:

"Now the country is going to destruction sure enough, and without remedy unless Congress interferes. A Milwaukee paper petrifies us with the astounding statement that the beer manufactured in that cream-faced town is terribly adulterated, and instead of being brewed from barley malt and hops, it is cheaply constructed from corn and rice. The tale is too horrible for belief! If the people of America can't get good and pure beer to drink, what is the use of living under a republican form of government and maintaining greenbacks at par with gold? A congressional investigating committee should be dispatched to Milwaukee at the earliest possible moment."

This book will help anyone who wants to de-mystify, understand, and form a closer bond with the most important ingredient in beer. John says that his principal reason for writing the book was to learn more about malt himself. He admits that the first brewery he worked at in Boston used 100 percent imported English malt but he wasn't exactly sure why. In the intervening years and through this book, he has learned why and shares his hard-earned experience with us. Some brewers simply scale up their use of malt from their homebrewing days, and others learn as much as they can so they can make educated choices when Mother Nature changes or destroys crops, or new ingredients become available, or unique flavors and colors of new products need to be created for a specific style.

Even though I am considerably older and have spent more years studying malt than John, I learned much by reading this very well researched book. You will too.

Dave Thomas
Beer Sleuth LLC
Golden, Colorado

Introduction

As I first thought about how to structure this book, I thought the starting place would be obvious. In Lewis Carroll's *Alice in Wonderland,* the King instructed Alice to "begin at the beginning and go on till you come to the end: then stop." But where precisely is the beginning in the story of malt? Does it begin with simple chemistry, or with the first historical records of barley farming? For this subject, one that I am so passionate about, perhaps the most appropriate starting point is where malt first entered my life.

Oddly enough, my relationship with malt began with my grandmother. She always had a diverse selection of sweets available for well-behaved grandchildren. In addition to the cellophane-wrapped, translucent yellow butterscotch candies that I would later associate with diacetyl, she'd occasionally have a dish full of chocolate malt balls. As I bit into the rich milk chocolate, my teeth would uncover a wholly different texture and flavor hidden within. Although I loved the sweet chocolate covering, the inner globe revealed a rich flavor that was not oily or fatty. It was sweet, but the sweetness did not come in a rush; instead it was long and drawn out, mixed in with elements of bread and cereal grain. As they say in music, it had sustain.

The oppressive August heat in Rhode Island came from the combination of ample sun and windless days in close proximity to the ocean. Although a dip in the cool ocean gave needed relief when the days got too sweltering, occasionally the family made an expedition to the Newport Creamery for a cooling ice cream treat. Their specialty, the "Awful Awful" (awful big and awful good), made it into semi-regular dietary rotation. The flavors in the malted milkshake echoed those malted milk balls from my grandmother's house. Those flavors were, quite simply, delicious and unique to my youthful palate.

My father has always been an adventurous beer drinker. His varied tastes and my beer can collecting hobby formed a nearly perfect symbiotic relationship. When beer transitioned from something that my older relatives drank to something that I was finally able to partake of, I do not recall a particularly malty flavor in the abundant but cheap beer that my peers were drinking. However, in contrast, the beers in my father's fridge struck me with their grainy complexity.

As I reached the age of independence, I moved into a group house with good friends who all happened to be employed in the food and beverage industry. My roommates included a classically trained chef, a food and beverage manager at a fine hotel, and a shift manager at the bar in the basement of the Hampshire House in Boston (the inspiration for the TV series "Cheers"). Together, we eagerly sought out whatever different beers we could find. The shelves of package stores throughout the greater Boston area during the 1980s yielded a wide variety of obscure beers from around the world. As we drank them in the name of research, we took copious notes on their individual flavor characteristics. Although we were unsure of the origin of many of the flavors, we knew what we liked.* Our home-built keg fridge was constantly stocked with delicious imported ales. The offerings from craft brewing pioneers like Sierra Nevada and Anchor Steam also came into the rotation as they became available in our region. The beers were dramatically unlike the insipid and nearly colorless light American lagers that dominated the beer aisles in those days. These beers had substantial malt-derived color, flavor, and body, and we loved to drink them.

This substantial interest (some would call it obsession) was a large factor in how I ended up both homebrewing and then working (first in the kitchen and eventually the brewhouse) at Boston's newly opened Commonwealth Brewery. The English styles made at the brewery were dominated by malt, and these flavors made their way seamlessly into the restaurant's food as well. The grain would go into soups, the wort into sauces, the beer into marinades, steamed mussels, and even sometimes desserts. Hops, despite

---

* It was many years later during the formalized flavor training that was an integral part of my Siebel education that I found out that the "German flavor" we often noted was identified with a proper name; "severe oxidation."

their potency in flavor and aroma, were not doing the heavy lifting; that work was left to the malt.

And heavy lifting it was; as an apprentice brewer I became intimately familiar with malt in that poorly engineered and inefficient brewery. We used 100 percent English malts; full container shipments would periodically arrive and need to be manually unloaded into off-site storage at the brewery farm, brought to Boston and winched to the second floor grain loft, and after milling, wheeled halfway around the block to the grist case. After brewing, I lugged the hot, wet bags of spent grain on my back as I navigated the broken and aging slate steps out of the basement.

We used different types of malt in the grist formulations for the various porter, bitter, and stout brews. These malts all looked, smelled, and tasted differently. I knew then that they all started as the same barley and somehow were transformed into the wide range of colors, flavors, and textures that drove and formed the distinct beers. But I did not know how or why. It was my assumption that given enough time, and enough experience in the brewhouse, I would eventually learn all there was to know about malt.

A thirst for brewing knowledge led me to Siebel Institute. After three years of working in (and eventually running) the brewery at Commonwealth, I was primed and ready. My class included experienced students from larger breweries all around the world, as well as a few somewhat clueless microbrewers.

I visited the Schreier malting facility* in Sheboygan Wisconsin with this class. A few of us came armed with what we were told was the best auditing tool in the brewer's arsenal: spongy white bread. Earlier in the week, one of our lecturers made a point of stressing the necessity of proper sanitation. We were told that bread, wiped on any surface in a well-run malting operation, should be appetizing enough to eat. We menacingly brandished the bread at our gracious hosts. While it was fun and telling of the quality of the operation, I am relieved that no maltster has ever tried this tactic with me at the brewery.

Although the scale of the operation seemed huge, with vast piles of grain soaking, growing, and being dried, the most memorable part of the visit were the aromas. Raw barley smelled dusty and dry. The bleach used to clean and sanitize stung our nostrils. A handful of growing barley fresh off the germination floor had a bright and clean smell–strongly

---

* This malt plant has been owned and operated by Cargill since it was acquired in 1998.

reminiscent of cucumber–while malt in the final curing stages on the lower deck of the kiln smelled rich and biscuit-like.

I took the reins at the Old Dominion Brewing Company in Virginia in 1991, and was given the opportunity to formulate and brew Continental type lagers. These Continental styles needed different malts than I had previously used in Boston. The American base malts performed differently in the brewery and contributed distinctive flavor characteristics to the beer. Brewing with the palest of American malts at Old Dominion made beer that allowed yeast or hops flavors to become the unrivaled star of the show. In contrast the pale ale malt I had used at Commonwealth was always evident in the background, never letting the other ingredients have their turn on the stage.

Old Dominion was making beer on a larger scale, and as the brewery grew, moving and processing the malt grew to greater prominence as well. Over time, the dumping of 50 pound bags of pre-milled malt by hand was replaced with larger storage, mechanical conveying, and automated weighing equipment. There was a corresponding increase in knowledge with each phase of growth. Some knowledge came from careful reading or inspired discussion, while other lessons were learned on the fly, from moments of near catastrophe.

In 2001, I made the move to Bell's Brewery in Kalamazoo, Michigan. In the time since, Bell's, like many other craft breweries, has experienced significant growth. As we have grown, so too has our knowledge about, and investment in, our various supply chains. The beer landscape is shifting, and as large breweries consolidate and pull away from public dialogue, we feel it is imperative that growing breweries support vital research related to raw materials. Homebrewers and craft brewers have different needs than those that have supported and directed these efforts up until now, and if we desire to continue to make better beer, it is imperative that we understand the nuanced challenges that our suppliers face.

Bell's has been actively involved with the American Malting Barley Association (AMBA) for a number of years, and has seen the organization slowly but steadily shift towards craft brewing. AMBA advocates for barley; coordinating the efforts of breweries and malting companies for research and government liaison. Through AMBA, the member companies collectively provide a unified message to farmers about what varieties should be grown, aggregate and apportion funding, and provide vital direction to barley breeders and researchers. It is a remarkably effective organization that strives to build consensus within the disparate stakeholders that make up its membership.

It was because of the work with AMBA that Bell's decided to grow barley in Michigan. Historically, central Michigan produced significant quantities of malting barley, but the rise of corn and soybeans as commodity crops had all but wiped that out by 2000. Our idea to farm came primarily from an inquisitive attitude. We felt that by purchasing a farm and growing barley, we might better understand the sometimes harsh realities that our suppliers deal with every day. Planting and harvest decisions, disease factors, and variety selection become far less abstract when actually standing in a field. Our philosophy of learning by doing has not always met with success, and we are better brewers because of this.

As of 2013, we grow about 400,000 pounds of barley annually. Over years of closely working with our malting partners, we have had the opportunity to gain deep insights about barley and malt. The experience of complete involvement in this full chain of stewardship: from earth to barley, barley to malt, malt to wort, wort to beer, and finally beer to enjoyment has been simultaneously overwhelming, fulfilling, frustrating, and transcendent.

It seems that as the brewery continues to evolve, so too does our relationship with malt, and we continue to look more closely at the starchy puzzle. It is the search for greater understanding that carries us through this book, from an malt-inspired overland expedition through Ethiopia in the time of Ras Tafari, and all the way back to high school chemistry class. Dig deep enough, and the story of malt takes us from the very foundation of human civilization to cutting-edge genetic engineering.

About This Book

We begin with the story of a fascinating character, Harry Harlan, who built the foundations for barley research in the United States. Before I started researching this book, I had never heard of him, and thus had no idea of how wide his influence was to barley and malt.

Chapter 2 focuses on how brewers use malt. Although the calculations needed to successfully achieve target wort strength and color are detailed throughout, there is more to formulating a great beer than just arriving at a number. The nuanced and delicate interplay of malt simply can't be reduced to a few data points. This chapter reveals the way that some imaginative brewers tackle the process.

The techniques and tools used to transform raw grain into a material that provides the flavor, color, and nutrients needed to make beer have evolved with civilization. Beer (and by default malt) has been a prominent and important component in many cultures. Chapter 3 gives an overview of that history, with some colorful interludes.

Emerging rootlets on germinating barley.

Just like brewing, practical malting can be as simple or complex as the artisan desires, but consists of three distinct phases; steeping, germination, and kilning. Steeping raises the moisture content of the grain and the barley responds as it would when moistened by warm spring rains; it starts to grow. As the grain germinates and the shoot and roots of the plant start to emerge, it is transformed internally. Since the "chit" (the emergent roots) would otherwise grow together during this "flooring" phase, the grain must be turned periodically. After germination is complete, the malt is kilned. This step both halts growth and develops characteristic flavors by drying the grain. Initially, the still wet "green malt" gives up moisture easily to the warm air that is blown through it, and it "withers." In the second part of drying, higher temperatures are needed to finish, or "cure," the malt. By the end of the malting process the very hard barley starch has been "modified" by breaking down its internal protein matrix and making it friable (easily breakable). Chapter 4 explores how malt is made, and the complex biochemical and physiological changes that occur during this process.

Although the vast majority of grain that is malted is pale colored with light flavors, by varying malting conditions, utilizing additional processing steps, or even starting with other grains, a wide range of specialty malts can be produced. Chapter 5 covers the production and flavor contributions in the five broad classes of specialty malts: high dried, caramel, drum roasted, alternate grains, and special processes.

Chapter 6 covers the chemistry associated with malt. Starch and protein are the main components of barley, and both undergo significant degradation and modification during the processes of malting and brewing. Just as large starch molecules are broken down into smaller sugar molecules, so too are proteins degraded into smaller polypeptides, peptides, and amino acids. The complex Maillard reactions that occur when amino acids and sugars are heated together create an incredible diversity of flavor and color that are an integral part of wort and beer.

The diversity of malt types is the subject of Chapter 7. Variation in ingredients and process allows maltsters to produce a wide range of malts. Classifying malt into individual types is an exercise similar to classifying beers into styles; there is a difference between amber and brown but where the line is drawn is an individual choice. It is useful for the brewer to recognize the broad flavor and functional attributes that different types of malt contribute to beer.

High quality barley is a prerequisite for the creation of great malt. The barley kernel is a complex organism with distinct anatomy. Like an egg, barley is composed of a protective shell (the husk), an embryo, and energy reserves in the form of starchy endosperm. These various parts contribute different elements during the mashing and brewing process. Beer quality is ultimately dependent on barley quality, so better brewers have gained knowledge of the challenges and opportunities of barley farming. The phrase "No Barley, No Beer" could be revised to "Know Barley, Know Beer" in Chapter 8.

There are many varieties of barley. Although they all can be turned into malt, they still technically different. Attributes like protein content are variety dependent, and thus variety has a large influence on the malt that is ultimately produced. Many brewers favor particular barley varieties for their beer, like the popular Maris Otter. Chapter 9 examines barley varieties in detail. Breeding and development of barley has provided maltsters and brewers with the best raw material possible, even as priorities have shifted over time. I feel that variety is an often overlooked aspect of malt and is necessary to understanding how it will perform in the brewery, and eventually, beer.

Malt analysis gives the brewer insight into what to expect, and allows her to make adjustments in the brewery to produce consistent results. Chapter 10 covers how to read and interpret a Certificate of Analysis (COA). By this point in the book, elements such as diastatic power (DP, a measure of enzymatic potential) and Free Amino Nitrogen (FAN, which quantifies soluble protein), will be familiar to the reader. Potential extract (the amount of content from malt that becomes dissolved in wort) is measured with the density scales of degrees Plato (°P) or by Specific Gravity (SG), and is important information for the brewer who wants to brew consistent beer.

Chapter 11 covers malt handling from a commercial brewery perspective. The theory, practice, and equipment used for malt milling are the subject of Chapter 12. Sprinkled throughout this book are visits to a few malthouses, both big and small, to get a sense of the variety of approaches to making malt. These mini malthouse tours give the reader some practical grounding, showing how this process is still alive and well in many places and on many scales, employing skilled malsters who all brewers rely on to supply enough grain to make their daily bread.

When I decided to write this book I had a simple goal; learn more about malt. Over my many years of brewing I had certainly gained much knowledge on the subject, but never studied it at a very deep, comprehensive level. This book gave me a reason to do just that. A lifelong passion for learning is nurtured by many influences. I hope that my efforts, and this book, spark a similar quest for knowledge in my readers.

1

Harry Harlan - the 'Indiana Jones' of Barley

A strange and winding path led me to the basement repository of the Kalamazoo Public Library to read about the Ethiopian people. Physically, the journey was just a quick walk down a set of stairs, but intellectually the trail was long and complex, sown with seeds of barley. I was in that book-lined basement on a mission to locate the 1925 *National Geographic* article "A Caravan Journey through Abyssinia" written by Harry Harlan. Harry had become a bit of an obsession for me, and I was trying to find out as much about him and his life as I could.

Unlike his fellow plant scientists, Harlan did not spend his 40-year career in a lab or writing reports, instead devoting himself almost solely to the practical study of barley. He documented the origin and genetics of the grain during his extensive travels, amassing an extensive library of the world's barley strains. His tremendous agricultural and genetic contributions in strain development and crop quality laid the ground-work for the varieties used by brewers and maltsters today.

Born in rural, western Illinois in 1882, Harry Vaughn Harlan graduated from Kansas State University in 1904. He took a job with the USDA and immediately departed for a three year stint in the Philippines that included opportunities to travel and experience the culture of Southeast Asia. The danger and adventure of life abroad prepared him well for what, at first glance, might appear to be the mundane world of plant breeding. Upon returning to Kansas to start his graduate work, he received a new assignment, and was sent to rural Peru to study barley for four months in 1913 and 1914. Under the South American sun, Harlan studied and collected local Peruvian barley types, marking the first (but definitely not the last) time he would travel far away from Kansas in pursuit of barley.

At the close of World War I, food reserves across war-torn Europe had begun to dwindle. In response, future president Herbert Hoover mobilized a unified relief effort aptly named the American Relief Administration (ARA). Harlan joined the ARA as a grain specialist and sailed for England shortly after the hostilities in mainland Europe had ceased. As part of the campaign, Harlan and his team were ordered to perform a sweeping, thorough analysis of the quality and availability of grain in post-war countries. An on-the-ground assessment of the situation would prove vital to the humanitarian agricultural mission that would eventually feed more than ten million people daily.

After crossing the English Channel and arriving in Paris, Harlan's team purchased a pair of new Cadillacs and began their journey southward to assess cereal grain crops. They crossed the Alps near Turin, continuing across northern Italy to Slovenia, Croatia, modern-day Serbia, Romania, and Ukraine, eventually turning back north towards Poland. While in Warsaw, Harlan traveled a few hundred miles east to assess additional grain fields. It was not until he was back in safe territory that he realized the surveying team had accidentally spent four days behind the Russian front lines during the Bolshevik Revolution. The disorder and chaos that war had brought was evident at every turn. In the course of his 5000 mile, 46 day car trip during the summer of 1919, Harlan had some of the most memorable eating and sleeping experiences of his life.

What lay ahead was never obvious; once he found himself hosted by royalty in an opulent and undamaged palace, and then, not a few days later, in an area so looted of food that he was embarrassed to take the meager bread from the hungry locals. Despite traversing horrifying, still-fresh battlefields, and passing untold numbers of displaced refugees, Harlan's keen eye was ever-trained on fields of barley. He assessed that the land race varieties he observed in Europe were substantially unchanged since the time of the Romans. The information he collected for his primary mission assured food security in post-war Eastern Europe for many

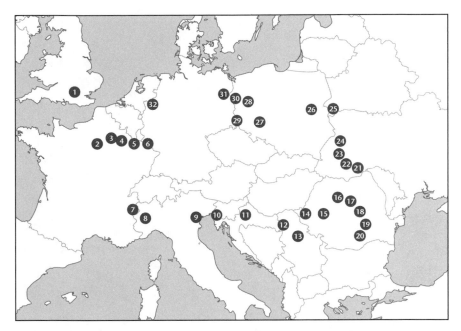

Fig. 1.1: Map created from locations listed during his post WWI travels in Europe.

1. London, United Kingdom
2. Paris, France
3. Fismes, France
4. Reims, France
5. Verdun, France
6. Ham-sous-Varsberg, France
7. Lanslebourg-Mont-Cenis, France
8. Turin, Italy
9. Venice, Italy
10. Trieste, Italy
11. Zagreb, Croatia
12. Ilok, Croatia
13. Belgrade, Serbia
14. Timisoara, Romania
15. Deva, Romania
16. Turda, Romania
17. Sighisoara, Romania
18. Brasov, Romania
19. Ploiesti, Romania
20. Bucharest, Romania
21. Chernivtsi, Ukraine
22. Kolomyya, Ukraine
23. Bukovyna, Ukraine
24. Lviv, Ukraine
25. Brest, Belarus
26. Warsaw, Poland
27. Wroclaw, Poland
28. Swiebodzin, Poland
29. Görlitz, Germany
30. Frankfurt an der Oder, Germany
31. Berlin, Germany
32. Kempen, Germany

years to come, and offered him insights about variation in local barley populations that influenced him for the remainder of his professional life.

In 1920, as a break from the Parisian nightclub scene during the ongoing post war negotiations, Harlan took an excursion to the French countryside with several friends. On this trip he met a fellow barley botanist named

Harlan in 1923 just before going into Ethiopia. Courtesy of the University of Illinois Archives, Jack R. & Harry V. Harlan Papers, RS 8/6/25

Mary Martini. This chance meeting meant more than both Harlan and Martini realized; she would become a lifelong collaborator and great friend to Harlan during his adventures in the world of barley. Together they bred, grew, and assessed new varieties in the US for many years, helping to create the scientific basis for modern barley variety development.

Many agronomists theorized that barley first evolved in Ethiopia, making it the perfect place for Harlan to continue his research. The country's complex political history involves periods when visitors were openly welcomed and others when visitors were seen as distrusted outsiders. Knowing he may be turned away, Harlan traveled to Addis Ababa in 1923. He knew that he needed to secure permission and safe passage across the infrequently traversed route to the Sudan. This special permission required a visit to the powerful leader known as Ras (head or ruler) Tafari (feared or respected). Officially named Haile Selassie, the young monarch (and heir to a dynasty that traced its origins to King Solomon and the Queen of Sheba) was an intelligent and internationally respected ruler. In the presence of this demi-god of a man (viewed as the returned Messiah by the well-known Rastafarian worshipers in Jamaica), Harlan presented the plans for his journey across the escarpment and plateau.

The expedition would cross the rugged and remote terrain that is the birthplace of barley. The route and length of the trip necessitated a large party of mules, porters, and guides. By design, Harlan chose both Coptic Christians and Muslims in an effort to increase the chances that he would experience the trip from multiple viewpoints. This approach had both benefits and drawbacks. "The Christians . . . not only drank, but not too playfully chased one another with razor-edged hunting knives and tried to throw one another over precipices. Some nights a surprising proportion of the Caravan was tied up."[1] Much to the surprise of Harlan (and presumably the delight of his financiers) the entire, elaborate party of twenty-plus men and forty mules cost only fifteen dollars per day.

Harlan was the first person of European descent that many of the residents of the areas the party journeyed through had ever seen. His account of the trip includes episodes of epic hospitality; skilled exhibitions of martial horsemanship followed by lavish banquets. Harlan recounts his meeting with Ras Kassa, as the night got too dark to continue their journey, "to light our way, some 600 followers of our afternoon's host lighted torches and streamed ahead and beside us as we rode down into the canyon and up the other side. The scene in its feudal setting had a beauty and a thrill that I am afraid I am not competent to transmit. As we passed through the gateway of each encircling stockade we found a section of Kassa's army drawn up in our honor. The wooden palace at the crest of the hill was no marvel of architecture, but it was pervaded by the dignity of a man who rules absolutely without tyranny and by the grace of a host who planned a dinner to please us rather than himself."[2]

Whether jostling with 30,000 devoted pilgrims during a Christmas celebration complete with fantastic costumes, dining on raw meat, or encountering invalids and lepers, Harry lived free and wild as he slowly journeyed to the head of the Blue Nile. But in the midst of all his adventure, he never forgot his primary quest, and along the way sought out and collected samples of native barley and other grains to bring back to the United States. Throughout his accounts a reader can sense the joy that Harry found in travel and exploration. Hardships like contracting typhus from ravenous bedbug populations or the imminent threat of attack by armed bandits are mentioned without injecting drama. To Harlan, using cotton and salt as currency, and the (temporary) theft of the steel box containing photographic equipment and letters of safe passage were all part of the tapestry of his exotic journey.

At the end of the fifty-nine day trek in 1923, when he had finally reached Gallabat on the border of Ethiopia and Sudan, Harlan's heart sank at the sight of telegraph lines (and in turn modern civilization), which signaled that his caravanning trip was over. He departed from the Sudan downriver to Cairo (via the Nile), and then across North Africa. No matter where the trip took him or how he got there, Harlan continued to collect and catalog the local barley.

As Harlan amassed samples, he identified the types and began the process of determining how varieties evolved and spread across the world. The incredible diversity of varieties and attributes in the Ethiopian highlands was quite notable. As he traveled through Spain later that year, it was evident that although the Spanish had brought barley across the Atlantic to the new world, the type they brought did not originate in Spain. The "Bay" barley grown in California had, in fact, originated in North Africa. When Harry found a Spanish variety that was well adapted to semi-arid conditions, it was his supposition that "it would make a good showing if seeded in the central part of west Texas."[3] It is unclear if this variety was ever actually tested in Texas, but by collecting it Harlan was able to make the genetic stock available for the USDA breeding program. His work provided the US government with enough raw material to properly research barley, which would be very important given the scientific progress of the mid-20th century.

In the spring of 1900, three independent researchers separately published work that rediscovered Gregor Mendel's "Laws of Inheritance." It had taken a full 35 years for the pioneering geneticist's work to become generally accepted by the scientific community, and the acknowledgement of its legitimacy spurred a rapid increase in applied plant and animal breeding. Harlan was able to witness firsthand the widespread adoption of these techniques for selective plant improvement. But buoyed by scientific validation or not, practical plant breeding has always been a numbers game based on random statistical chance. The applied science of field crossing that Harlan practiced made little use of Mendel's theories. The more crosses made, the deeper the genetic pool, Harlan surmised, regardless of dominant or recessive genes.

At the practical level, the number of genetically expressed traits and the non-obvious interactions between them created the need for significant fieldwork. At one point, Harlan made every possible cross between 28 distinct barley varieties. The practical management of the 378 distinct combinations was so daunting that it prompted him to quip that anyone attempting such a task "should have an island to himself until the process is over. He is not fit to associate with nice people."[4]

By 1950, as a result of his trips (such as the 1923 float down the Jhelum River through present day Pakistan on multilevel houseboats complete with silk tents), Harlan had more than 5000 different barley varieties at his disposal, collected from every corner of the globe. Although his travels include jaunts to China and Japan and surveys of remote high country Peru, he always returned to the fields of Aberdeen, Idaho, and Sacaton, Arizona, for the practical work of applied plant breeding.

Harlan found very wide variation in the barley populations collected from around the world. He investigated the differences between two-rowed and six-rowed, naked and husked, spring and winter barleys, and traits specific to certain cultivars, such as straw length and dormancy. Although breeding for increased agronomics was the main focus, interesting crosses between "freak" barleys were also attempted, often resulting in intriguing and novel characteristics that he and Mary Martini watched grow with curious delight.

From years of working in the adjacent fields, Harlan grew to love the close community of Aberdeen, Idaho. He wrote with a clear fondness of the colorful collection of characters who resided in the tiny town with a population of 1,496.[5] When the workload permitted, Harlan would take visiting USDA researchers fishing in the nearby Sawtooth Mountains.

If it was still possible, I would relish the chance to directly experience the tremendous hospitality and see firsthand the valuable research that the combined forces of Harry Harlan and Mary Martini regularly delivered. The samples they collected have been repeatedly grown, characterized, tested, and cross bred, and together they built the diverse genetic foundation that resulted in the development of many publically available varieties used for feed and brewing. Somehow the thought of a day of fishing in the mountains followed by an evening of convivial storytelling with this globetrotting barley researcher seems difficult to top.

References

1. Harry V. Harlan, "A Caravan Journey through Abyssinia", *National Geographic*, Volume XLVII, No. 6 June, 1925, 624.

2. Harry V. Harlan, *One Man's Life with Barley*. (New York: Exposition Press, 1957) 45.

3. Harry V. Harlan, *One Man's Life with Barley*. (New York: Exposition Press, 1957) 37.

4. Harry V. Harlan, *One Man's Life with Barley*. (New York: Exposition Press, 1957) 98.

5. Riley Moffatt, *Population History of Western U.S. Cities & Towns, 1850-1990*. (Lanham: Scarecrow, 1996) 90.

2

Malt: The Soul of Beer

Like stock creates the base of every great soup, malt provides several key attributes that define beer as we know it, including color, flavor, body, and, eventually through fermentation, alcohol. When formulating the malt bill for a beer, a brewer should take each of these factors into consideration. Grain bills vary widely; some may utilize only one type of malt while others are complex combinations and mixtures of several types. This chapter will explore the innate diversity of malt, how it is used in the brewhouse, and how other professional brewers utilize different types to create distinctive, balanced, and flavorful beers.

Flavor

Throughout my years of brewing with Bell's, Old Dominion, and Commonwealth Brewing, I had many discussions about malt and its contribution to beer flavor. Beer drinkers often have differing opinions as to what defines "malty" flavor. As brewers and maltsters, the techniques we use to capture and produce those flavors also vary. Our exposure to other flavors and our subjective perceptions impact the way we think malt manifests in the finished product.

If I were to stick a pin on a map to define what malty means to me, it would land close to Munich malt. Although malt also contains toasty, sweet, burnt, and husky flavors, the rich, aromatic flavor of Munich is what springs to mind when someone says the word "malt." Even small additions of Munich to a recipe seem to fill out the middle palate. If I am writing a recipe, chances are that there is some Munich malt included. Beyond that, I'm very open to malt styles; German, Belgian, English, and lots of American malts all get their turn on the mashing stage.

Chewing malt is a vital part of beer formulation; it is the best way to explore and analyze the combination of subtle differences between different varieties. It's amazing how many people are disconnected from their senses of taste and aroma, even those within the industry. Beer enthusiasts and brewery employees seem reluctant to actually eat ingredients during tours and trainings, and it takes considerable prodding to get people to actually put raw materials in their mouths to truly experience a flavor. Seeing, hearing, and feeling are wonderful sensations, but when it comes to beer and malt, those senses kind of miss the point. Munching on malt lets a person assess more than just flavor, giving a brewer a direct example of crucial quality metrics such as differences in friability[*] and moisture content. There are many practical and delicious reasons to put malt in your mouth, all of which help decide what grains will be best for your beer.

Formulating a Grain Bill

Several factors need to be weighed and balanced when formulating a grain bill. First, a brewer should consider how much fermentable extract will be present in the wort, as this will directly relate to the alcohol content of the final beer. The next factor is unfermentable extract, which provides the desired flavor and body in beer, and mainly consists of larger carbohydrate molecules. These longer chain sugars will persist through fermentation, ending up in the finished beer, contributing important characteristics because they could not be consumed by the yeast. The ratio of fermented to non-fermented sugars affects the dryness of the beer; if all sugars are consumed, the beer will be dry, and any remaining sugars not eaten by the yeast will impart perceived sweetness to the beer. As described in Chapter 6, pound for pound, long chains sugars are not particularly sweet, but do contribute to the mouth-filling body desired in many beers.

The simplest grain bill consists of a single type of malt. This single malt needs to contribute starches, enzymes that can deconstruct said starches

---

[*] The property of being easily crumbled or pulverized. Fully modified malt should be friable, unmalted barley is not.

into fermentable sugars, Free Amino Nitrogen (FAN), trace minerals, and vitamins. All of these are necessary to create a wort capable of nourishing yeast as it transforms sugars into alcohol during fermentation. Unfortunately, not every type of malt meets these requirements. A higher temperature during kilning denatures the conversion enzymes in specialty malts, thus destroying their ability to transform starches into sugars. Any malt that has sufficient enzymes to convert the starches it contains is capable of being used as a base malt.

Base malts are pale in color and make up the majority of the grain bill in most beers. In addition to base malts, brewers may add other specialty malts or adjuncts to the grist to produce a wide range of flavors. Most specialty malts are produced by applying additional heat during the kilning stage at the malthouse or by using specialized roasting equipment. Adjuncts are non-malt sources of fermentable sugars. Although corn and rice are the most well-known, many other ingredients can and are used by breweries all over the world. Like malt, if the adjunct contains starch, a sufficient source of enzymes is needed to break it down. Some native starch adjuncts will also require additional cooking before being added to the base malt mash. Traditionally, well-modified pale malts provided the enzyme source required for adjunct conversion. For this reason, minimum enzymatic potential in malt has always been a critical quality metric. Today, commercial liquid-enzyme preparations can also be added to the mash to enhance enzyme activity.

Brewers aiming for a specific wort concentration begin grain bill formulation calculations by determining how much malt will be needed to hit this target. Most base malts have the potential to release roughly 80 percent of their dry mass into wort. Finished malt contains roughly 4 percent water, so 100g of malt "as is"[*] would contribute about 80 percent of 96 grams (or 76.8g) of extract under ideal conditions. Different malts, grains, and adjuncts vary in the quantity of potential extract and moisture, and by summing the calculated contributions of each element of the grist it is possible to predict the performance of the entire mash, although brewhouse efficiency also needs to be taken into account. Efficiencies vary widely and are influenced by brewhouse design and engineering, as well as mash profile. Although calculating these parameters is relatively simple through the use of apps or spreadsheets, an understanding of the underlying principles helps a brewer understand why the malt used matters, not just how.

---

[*] Malt measured "as is" includes the moisture. "Dry Basis" calculations assume all moisture has been removed.

At a basic level, the measure of wort strength can be simplified to "How much extract did I manage to get in the water?" During mashing, hot water, adjusted to a specific temperature, pulls soluble extract from the grain and in turn, changes the density of the water, creating wort. Liquid density can be measured by the scientific unit of Plato (°P), and the often used reference tables are based on sugar percentage of a solution. A 14°P solution results when 14 grams of sucrose are mixed with 86 grams of water. The volume that 100 grams of this solution occupies is less than the volume that 100 grams of water would fill, making it denser than water. At 60°P, the specific gravity (SG) (or the solution's density relative to water) can be roughly calculated by the formula:

$$SG=259/(259-°P)$$

A 14°P wort has a density about 1.057 times that of water. For water 1 gram (g) = 1 milliliter (ml) so a 100g solution would occupy 94.6 ml (100/1.057). If we wanted to make 94.6 ml of 14°P wort, we need to liberate, dissolve, and recover 14 grams of extract from the malt.

For calculation purposes, the individual components of malt can be considered in sections.

- Water (malt leaves the kiln at about 4 percent moisture but that value can vary by a few percentage points)
- Husk material and other insoluble carbohydrates
- Protein (only some of which is soluble)
- Enzymatically convertible carbohydrates (the source of fermentable and non-fermentable sugars)

When we separate the categories that dissolve into wort from those that cannot, we find that the gravity of wort is dependent on soluble proteins and carbohydrates.

As we will find in Chapter 10, the amount of potential extract (or the measure of soluble proteins and carbohydrates) can be expressed in several ways. For basic brewing calculations, the Coarse Grind, As Is (CGAI) test value helps a brewer decide what malt to use for a specific beer. For example, malt with a CGAI of 80 percent released 80g of soluble extract for every 100g of malt under laboratory conditions. As this value decreases, the amount of malt needed to meet a beer's specification needs to increase.

Brewhouse efficiency (the ability to actually recover potential extract from the grain) needs to be factored in when calculating a malt bill. Many factors can influence efficiency. As wort strength rises, efficiency inevitably declines. Prior performance of similar mashes is the best tool for determining future mash efficiency.

The extract determination steps in writing a recipe are:

1. Decide how strong the wort should be and the final volume
2. Compute the total extract needed in the wort, using these two values
3. Adjust the value of potential extract based on brewhouse efficiency
4. Calculate, sum, and readjust the relative extract contributions of various malts to match the needed potential extract

The extract tables published by the American Society of Brewing Chemists (ASBC) describe the relationship of °P, SG, pounds of extract per barrel, and kilograms of extract per hectoliter. The brewing department at Bell's uses degrees Plato, pounds, and gallons in most of its calculations. [*] Homebrewers are likely more familiar and comfortable with specific gravity "points per gallon." The specifics of the system don't really matter, as long as you actually have one.

A rough guide for conversion between SG and °P is that SG=[(°P *4)/1000]+1. A 12 °P wort has an SG of about 1.048. At higher concentration the correlation diverges but for more "normal" ranges it gets you close.

Let's use an example: If I wanted to make 10 barrels (bbl) of a 12°P, medium colored beer with a full malty flavor, I would start by sketching out a rough recipe based on percentage of malt types.

Based on research and experience the starting point for the grist formulation for my beer will be:

Malt	Grist Fraction
Pale Malt	80%
Munich Malt	12%
Crystal	7.5%
Black	0.5%

Next I calculate the total extract needed. By reviewing the ASBC tables we find that 12°P wort contains 32.45 pounds of extract per bbl. I thus need to deliver 324.57 pounds of extract to the brew (32.45 X 10=324.5). Based on

---

[*] Although kg and liters are easier, I do it all on a spreadsheet which makes the math easy.

prior experience I assume that this brewhouse has an efficiency of 90 percent, and therefore need a total of 360.6 pounds of potential extract in the grist (324.5 / .90 = 360.6)

Extract needed	32.45	lbs/bbl
Volume (in bbls)	10	bbls
Volume (in gallons)	310	gallons
Total extract needed	324.5	lbs
Brewhouse Efficiency	90%	
Potential extract required	360.6	lbs

The extract and color values as reported on the Certificate of Analysis (COA) for the malts are:

Malt	CGAI	SRM
Pale Malt	80%	2
Munich Malt	78%	10
Crystal	72%	35
Black	50%	500

I can determine the needed extract contribution from each malt by multiplying the grist percentage by the total extract available from the lauter tun (360.6 pounds):

Malt	Grist %	Extract Contribution
Pale Malt	80%	288.4
Munich Malt	12%	43.3
Crystal	7.5%	27.0
Black	0.5%	1.8
Total	100%	360.6

To calculate the grist weight needed of each malt I divide the extract contribution by the CGAI value.

	Extract Contribution	CGAI	Grist lbs
Pale Malt	288.4	80%	360.6
Munich Malt	43.3	78%	55.5
Crystal	27.0	72%	37.6
Black	1.8	50%	3.6
Total			457.2

I now have a rough grist bill for my brew and can next check the calculated wort color. This value is derived by multiplying the color contribution of each malt by the total weight used. The resulting values, with the confusing units of SRM•lbs, or malt color units (MCU), are then summed.

	Grist lbs	SRM	SRM•lbs
Pale Malt	360.6	2	721
Munich Malt	55.5	10	555
Crystal	37.6	35	1315
Black	3.6	500	1803
Total			4393

This total is then divided by the total number of gallons to determine the wort color.

4393 SRM•lbs divided by 310 gallons results in an approximate wort color of 14[*] SRM.[**]

By using these calculations, a brewer is able to produce a rough draft of the grist bill. At this point the recipe can be fine-tuned. By utilizing a spreadsheet a brewer can quickly evaluate the effect of subtle changes in weights and extract efficiency in a manner similar to a recording engineer using a mixing board. A brewer can now also make adjustments for ease of brewing by rounding most malt weights to full pound increments at this scale.

---

[*] Holle, Stephen, R., *A Handbook of Basic Brewing Calculations,* Master Brewers Association of the Americas, St. Paul, MN, 2003.

[**] It should be noted that the SRM color scale is based on, and is approximately equal to, the Lovibond scale, which is based on visual comparison to reference standards. The actual measurement of wort and beer color in SRM is defined as ten times the absorbance of light at 430 nm wavelength transmitted through a ½ inch diameter cell. See ASBC Methods of Analysis, Beer-10.

Quantifying Wort Fermentability

As most brewers are aware, the solution concentration of wort can be expressed in several ways. 900ml of water weighs exactly 900g. If we dissolve 100g of sugar into this water the mixture will have a total weight of 1000 grams but will only occupy 962ml; this solution is 10 percent sugar by weight and is denser than water by a factor of 1.040. We can describe this solution as 10°P (or degrees Balling) and having a specific gravity of 1.040. When sugar is fermented it is transformed into roughly equal amounts of alcohol and CO_2. Because alcohol is lighter than water, if all of the sugar were to ferment into alcohol the resultant solution would have a specific gravity of less than 1.000 and an "apparent extract" of less than 0°P. A typical beer made from 10°P wort might finish measuring 2.5°P. This beer has had 75 percent of its "apparent" extract depleted. This beer would have an Apparent Degree of Attenuation (ADA) of 75 percent.

Thus, original gravity, final gravity, original extract, apparent extract, real extract, alcohol by weight (ABW), alcohol by volume (ABV), and calories are interrelated and those relationships can be expressed mathematically.

As a technically-oriented brewer, I find that an awareness of soluble extract, enzyme packages, and entrained moisture is crucial to brewing quality beer. When a mastery of the scientific aspects meets the creative, artistic side of recipe formulation, world-class beers are born. For me, a "great" brewer has the ability to consistently make sublime beers that are full of character, even when the ingredients she works with are constantly changing due to the whims of Mother Nature and process variation. Without a deep understanding of the causes and potential compensations, our reactions to these challenges may not be as effective, and the consistency and quality of the beer would suffer as a result.

Maintaining consistency in brews begins at formulation. Identifying critical parameters with the aim of leaving some wiggle room to adjust down the road means a brewer has some flexibility if something should go wrong. Although extremely low protein and exceptionally low color malt may be available now, unless you are brewing a one-off beer, painting yourself into a very pale beer corner may be shortsighted, as the availability of that malt may change leaving you unable to produce the same beer.

Malt based analytical data is helpful, even critical, to successfully adjust a recipe to account for typical variation and availability of ingredients. Even small variations in malt can have dramatic effects later in the brewing process, so careful review of analytical data is crucial to maintaining the consistency of a routinely brewed beer. For example, in contrast to industrial brewers, at Bell's the beer is brewed to sales gravity, so post-kettle dilution-water additions are not part of the brewing program. Bell's adheres to tight specifications on wort strength and fermentability, in part because of the narrow range of deviation in declared alcohol percentage that is allowed for by regulators. If fermentable sugars change, so too will the alcohol level of the finished beer, so every step is taken to ensure all the data adds up before mashing ever begins. The selection of the types and quantities of grains used for brewing largely defines the beer that will be produced and is therefore one of the most important decisions that the brewer makes.

Color Calculations

Color prediction is another fundamental aspect of beer formulation. Lovibond, Standard Reference Method (SRM), and European Brewery Convention (EBC)[*] are all scales used to quantify color. The SRM value of the malt is roughly equal to the wort color achieved when 1 lb of malt is mashed in 1 gallon of water (see sidebar). Dark colored malts can have hundreds of times the color potential of pale malts. It should be noted that wort develops additional color in the kettle as a result of melanoidin production which we cover in Chapter 6. This is particularly significant for very pale beers; a 1 degree SRM color change in a porter would go nearly unnoticed.

The simple quantification of color does not tell the whole story. Analytically, a brilliant orange wort and muddled grayish brown wort may generate similar SRM color values, despite a vast difference between them. A complete discussion of hue, depth, and perceived color is beyond the scope of this book, but fortunately several excellent references are available on-line. The work of Bob Hansen of Briess Malt & Ingredients Company is a great place for further exploration.[**]

---

[*] For our purposes 1 Lovibond = 1 SRM = 2 EBC units.
[**] http://www.brewingwithbriess.com/Assets/Presentations/Briess_2008CBC_UnderstandingBeerColor.ppt

Advanced Malt Color

Standardized malt analysis is performed using a "Congress Mash" which consists of mashing ground malt with a specified quantity of water, filtering out the solids, and measuring the properties of the resultant wort. Three methods exist; the Institute of Brewing (IoB)*, European Brewery Convention (EBC), and the American Society of Brewing Chemists (ASBC). The IoB is the standard in the United Kingdom; the EBC in continental Europe; and the ASBC in North America. Although the methods differ slightly from one another, they are the defined standards that can be used to compare the qualities of malt samples.

The Congress Mash is made by mixing 50 grams of malt with 400 grams of water. This equates to a mash made at 1.04 lbs/gal (12.5 kg/hl or 32.3 lbs/bbl); which is weaker than typical mash/wort concentrations.

Wort and beer color is assessed and reported using one of 3 methods; Lovibond, SRM (Standard Reference Method) or EBC. The Lovibond method is the oldest and uses direct comparison of a sample to colored glass slides. The more modern SRM and EBC methods photometrically measure the light absorbance of filtered, clear samples at the light wavelength of 430 nanometers (deep blue).

SRM values were created to agree with the original Lovibond scale and are fundamentally interchangeable. EBC and SRM values differ by a factor of 1.97, making EBC roughly double that of SRM.

Malt Color Units (MCU)

Wort color can be roughly estimated by converting grain ingredients into malt color units (MCU). MCUs are determined by multiplying the number of pounds of each malt in a recipe by their respective color values (in SRM), and then dividing the sum of inputs by the number of gallons of wort produced.

The MCU method works reasonably well for wort with SRM values below 8, but above that it is not accurate as the SRM and EBC scales are logarithmic and the simple calculation is linear. Dan Morey, a competitive homebrewer in the Midwest and contributor to *Zymurgy* and *Brewing Techniques* magazine, published an equation to approximate color at higher ranges:

Beer color (in SRM) = 1.4922 * MCU 0.6859

Plugging MCU values into the above equation gives the following results:

MCU	1	2.5	5	10	25	50	100	250
SRM	1.5	2.8	4.5	7.2	13.6	21.8	35.1	65.9

Appendix B of John Palmer's book, *How to Brew,* has a comprehensive exploration of wort and beer color. I like his simplified color model of Beer Color (SRM) = 1.5*(wort color in MCU)^0.7 as it is certainly accurate enough for general formulation use.

Although a host of factors such as pH, boiling intensity, and brewhouse efficiency influence beer color, malt color is the primary driver. The greatest value of getting malt color values in a Certificate of Analysis (COA), a document issued with each malt lot listing product specification test results, is the ability to find process deviations and react to them before they manifest themselves in beer.

The spectrum of malts available today is truly impressive. The diversity of color, flavor, and functional properties on the market can be overwhelming, especially to new brewers. Successfully hitting a target style can involve mixtures of many different malts, so in addition to assessing malt by chewing, talk to fellow brewers for inspiration. Discussing what they feel is necessary for a given style, what they like (and what they don't), can help give perspective on what malts to use, or inspire a creative spin in a traditional grain bill.

Brewing Perspectives

There are a wide variety of philosophies and techniques used by brewers in grain bill formulation. Creating a complex yet well balanced beer is equal parts art and science. It is difficult to quantify the multitude of individual flavor contributions from specific malts, so the brewer must move beyond the spreadsheet. It can be very enlightening to learn how different brewers approach this challenge, and understand what they like and what they do not.

Conceptualizing, brewing, assessing, and tinkering with a recipe, or "massaging" it into shape, is how most brewers approach developing exceptional beers. The process of dialing in a recipe can sometimes take years, but that path can be shortened with sufficient vision, experience, and careful calculations. Brewers approach beer formulation from many different angles; some technical, thriving off spreadsheets and gross percentages; others research-based, scouring available resources to get a sense of the range for a given style. And then there are the extremely intuitive brewers who are able to bring a holistic sense of how the individual parts will add up and interact to form a beautiful and balanced beer. To use a golf analogy: these brewers seem to start already on the green. One measure of competence is performance in beer competitions. Great beers don't always win, but it seems that consistently winning beers generally taste pretty good. A sense of what goes into making a consistently winning beer can be found by talking with consistently winning brewers. Talking to great brewers about how they envision a beer and then pull the malt bill together reveals individual viewpoint and methodology, and gives insight into all brewing processes. Thoughts from a few such brewers are shared in the following section.

Wayne Wambles, Cigar City Brewing Co. (Tampa, FL)

Wayne Wambles is the main creative force behind the delightfully complex beers at Cigar City brewery in Tampa, Florida. On the continuum of technical focus to artistic vision, Wayne embodies the artistic approach. As we spoke, it became clear that his conceptualization takes many cues from painting. He described base malt as being the canvas that provides structure for beer. Mouthfeel is analogous to texture of the brush strokes. In his view, specialty grains provide color—toasted and caramel malts impart bright colors whereas dull colors come from darkly roasted malts. With this outlook it is not surprising to find that Wayne often builds resonating and powerful malt flavors in his beers through very complex grain bills. "With *Big Sound*, our Scottish beer, we use 11 different malts; mainly toasted and caramel malts."

As he works towards the initial recipe he may remove or add more malt to the list. He describes his approach to formulation in a simple way: "Generally I sit down and try to get an idea of what I want to make: I have a preconceived notion of color, gravity, and IBUs. Then I determine what type of malt flavors I want. After I write down the malts that I want to use, I then fill in the gaps with percentages and IBUs." It is only after the percentages are determined that he plugs the recipe into the old version of Pro Mash brewing software that he has been using for years. This gives him a good estimate of what he should expect in the brewhouse.

Here's how Wayne might formulate a theoretical English barley wine. To build the desired underlying supporting complexity he would actually start with two different types of Maris Otter base malt. His approach was that "when building a wine, great vintners will oftentimes use grapes from different parts of the land." Wayne would add some cara malts and a bit of Vienna or another toasted malt to the Maris Otter to complete the mixture.

Wayne likes to use the variety Maris Otter as a base malt in many different styles; for porters he feels that it builds a complex malt flavor that is characteristic of the style. He describes that flavor as malty-biscuity with a bit of earthiness. At 3°L it tends to add a lot of flavor without much color and is therefore a great foundation for malt forward and low alcohol beers. "It has built-in flavor that you just don't get with pale. If milled properly I think that it mashes better. If you just crack it, then it lauters very well." At Cigar City they mainly use Simpsons as their supplier for Maris Otter.

"I like English specialty malts a lot. American malts are very clean but the English cara and crystal malts seem to have more depth. They carry fruit flavors that American malts do not." For the bright clean malt flavors needed to complement the flavorful hops in American IPAs he prefers caramel malts from Briess or Great Western.

When asked about what other malts he likes to use, Wayne eagerly shared his love for the figgy and prune like flavors that English dark caramel malts such as Baird or Simpsons provide noting his appreciation for their contributions to styles such as Belgian Dubbel or Belgian Dark Strong. In the robust porter style he likes to use pale chocolate malts. "You can stack chocolate malts up and jack the percentages up and get a more rounded beer without too much dry, burnt, char flavor." He likes Carafoam malt to add body as well as Chocolate Rye because "you can use lower percentages to get great rye flavor without a difficult lauter." Briess Special Roast™ is another go-to malt in his recipe development. To elevate complexity Wayne recommends adding some Victory®, aromatic or biscuit, to the grist but warns, "if you overuse it, it turns into a mess."

He recognizes, and is happy to elaborate on, why formulating a complex beer requires a deft hand to prevent crossing the line from subtle to over-dominating. For example, when tasting Double IPAs he feels that a touch too much caramel can transform the beer from delightful to "a big sweet mess" that often tastes under-attenuated. Similarly he noted that excessive black patent malt leaves a final finish dominated by char and ash flavors. Too much toasted malt, like Victory, imparts a flavor like unsweetened peanut butter and tastes awful to him. It seems like all the liquid is removed from his mouth when there is too much brown malt in the beer.

Wayne is careful to not dilute the malt characteristics with unintended yeast flavors. He likens it to "covering a painting with wax paper" when beers are served with too much suspended yeast. With specialty malt additions that sometimes top 40 percent it is clear that Cigar City's commitment to substantial malt flavor is a key part of their brewing program.

Jen Talley, Auburn Alehouse (Auburn, CA)

Jennifer Talley's ebullient personality is a window into her passion for brewing. She has been brewing professionally for more than 20 years; first at Squatters Brewery in Salt Lake City, next at Craft Beer Alliance (Red Hook) in Seattle, Russian River Brewing Co., and finally at Auburn Alehouse. Her many years of pub brewing allowed her the opportunity to trial and refine both individual recipes as well as the processes for developing them. The desire to make flavorful beers within the lower alcohol constraints required by the state of Utah inspired her to brew creatively.

For Jen, recipe development begins with substantial research. "Before I sit down, I want to learn about the style; read, taste, talk to other brewers. Determine what I like and don't like. Know the history; learn from it, not to copy but to find what I like about it. Only then do I put pen to paper."

After determining her target parameters including wort starting gravity, alcohol percentage for finished beer, color, and most importantly malt flavor notes, she begins her calculations. The underlying foundation of the beer is the base malt. She feels it's important to actively taste the malt as the beer takes shape in her mind. Of particular importance are the specialty malts she needs to use to get to her final objective. She is also keenly aware that there is a synergy with other ingredients in the beer. "You can destroy everything you are trying to bring out from the malt by over, or under hopping, the beer. The malt does not exist in a vacuum. You need to think about how they all interact together."

When I asked her to lead me through the thought process she would use in the initial stages of formulating a beer that exemplifies great malt flavor, she began by succinctly describing the beer: "Mid-color, malt forward with middle range hops to provide counterpoint." She would start with about 40 percent Maris Otter pale ale or Gambrinus ESB malt and back that up with the house base malt, likely a standard American two row pale malt. To build a different layer of malty character she would add roughly 15 percent of 10 degrees Lovibond (°L) Munich malt. Adding 5 percent of a mid-range 40–60°L Cara Aroma would complete the initial recipe. Small corrections would likely be made after transferring these quantities into her formulation spreadsheet. The tweaks to the final recipe might include an addition of Carafa III of less than 1 percent to fine tune the color. When the first batch of the beer was in her glass she would eagerly seek opinions to determine what additional modifications would be beneficial. For her, the benefit to brewing at a pub was that ready access to different points of view; "I always have open ears to my customers." She considers the recipe a living document and that subtle changes are not just accepted, but desired.

When asked what malts stand out to her after years of brewing, she was quick to respond. "Weyermann Pilsner malt. You just cannot substitute; you need to spend the money on the great malt for the Pilsner style. That flavor is hard to describe, there is a slightly bready but bright maltiness to it."

She likes Cara Aroma from Weyermann and feels that it brings a depth of caramel maltiness with an added complexity. "It is gentle and bold at the same time with a great color range." Brewing in Utah taught her how to maximize flavors within the constraints she faced. "If you want a beer that has a malt background you only have so much room. You have a limited amount of malt that you can actually add, because you need fermentable sugars too."

"I love Hugh Baird roasted barley"; she feels it should be a leading star in a dry Irish stout. To increase color without much flavor she likes Carafa 3, the dehusked black. "No one will ever know it is in your beer. At 1–3 percent it can give you the black you want or the red you want to match the style." She feels that it is hard to get true red hues, as opposed to copper or amber and visually there is a large and important difference.

For Jennifer, the largest problem she encounters as she tries new beers is an excessively heavy hand with specialty malts. "Someone has mistaken punching you in the face for flavor. Overuse is inarticulate in formulation.

When I just can't finish my pint, typically they have overused specialty malt." In her experience key culprits include Victory®, chocolate, Carafa III, and even roasted barley. "I have a hard time with biscuit. It is easy to overdo it."

When asked about her least favorite malts, she replied, "Peat malt; if you want smoke flavor use real smoke malt. I don't like peat malt, it is too phenolic. You can do better with a real smoked malt." It was interesting to hear that Wayne Wambles of Cigar City holds the same opinion "Anything peated," Wayne says, "I have trouble with it in any beer; it is my number one hated malt."

Jon Cutler, Piece Brewery & Pizzeria (Chicago, IL)

Chicago's busy and cool Wicker Park neighborhood is the home of Piece Brewery & Pizzeria. Jonathan Cutler has headed up the brewing operations since they first opened in 2001. In the time since, the beer produced in the tight brewhouse has garnered plenty of local acclaim, and quite a few medals at the World Beer Cup and Great American Beer Festival.

When Jon Cutler formulates the malt bill for a beer he thinks about it in three parts. "Base malt is the foundation you build on; with it you're probably already 90 percent of the way there. I'm always thinking of base malt first." German Pils malt is his obvious choice if he is working on a German style. If the beer is American then he will use two-row pale malt. He uses the second portion of malt to "turn the dial and point to the style." He changes the base malt slightly by building one malt on top of another. The addition of malts like Munich or crystal to the base malt adds a desired complexity to match the style. He uses the third part, his finishing malts, to tweak the beer's functionality. "It serves a purpose; enhancing head or bringing up the color. I'll use oats, wheat or maybe a bit of Carafa™ to bring it all together."

For Jon the composition of a beer is like a piece of music. "It's lyrical, like a song. Your base malt is kind of your bass track; it's your rhythm, the structure the whole beer is built on. Then you bring in some guitar or keyboard, something interesting. Then you come in on top of your vocals and tie it all together, put a bow on it; some little thing that makes it. It turns a boring song into something that you say, 'that's kind of different. I haven't heard that before.' The song needs to have a hook." He adds that, "your first batch is like a demo tape. You may add some backing vocals if need be. It might need a little more cowbell. The malts must work together mellifluously before you know that the cut is ready to press."

Jon's favorite base malts are the pale and two row malts from Rahr, Weyermann's Pils, and wheat. Common instruments for middle section? "I am a fan of Munich, English pale, Melanoidin and both light caramel like C-15, and C-60." He uses a wide variety of finishing malts; Carapils or dextrin are essential to his beers. "There is nothing else like them." He is a fan of the whole Weyermann catalogue and loves the unique flavors of the Cara–malts (Foam, Hell & Wheat).

"The variety of roast malts available is incredible. What kind of roast malt for your stout? Different malts give very different results." He loves the counterpoint that Munich or chocolate malt brings to a stout. By sitting in the middle, "They tie it all together, it doesn't just go from base to char."

Jon's least favorite malt is C-60, but he also recognizes that it is essential to craft brewing. "It has great character and is the quintessential malt for American Pale Ale but is prone to oxidation within a very short period of time. There is nothing worse than the ribey[*], oxidized C-60 flavor. It's the perfect middle ground malt but it is a catch-22. It can make a perfect beer but it can also break it. You can only have it for so long before it comes back and bites you in the ass."

His parting advice? "Learn by doing. Sometimes when you taste a malt, it doesn't necessarily translate to the beer. You need to taste it in the finished beer and learn from that. You are now playing/brewing it live and waiting for your audience/customer feedback. I don't worry about the critics. I'm my harshest critic and I know what it should sound/taste like to me. Screw RateBeer, and the like. If I put my blood, sweat and tears into it, then it is what I wanted it to be."

Bill Wamby, Redwood Lodge Brewery (Flint Township, MI)

Bill Wamby has demonstrated his brewing skills by winning a series of Great American Beer Festival medals for Michigan's Redwood Lodge. His meticulous approach to the competition starts with grain bill formulation. A careful reading of the category description acts as a roadmap, which he navigates to build his recipe. When it comes to malt selection what does he look for? "Uniform, plump kernels. I don't want to see

---

[*] Ribes is a British flavor descriptor for bruised tomato or blackcurrant leaves. It is associated with catty flavor.

'thins'," and for that reason he is particularly fond of English malts. Another favorite of his are the Moldavian malts. "Certain malts just have a stronger depth of character."

As Bill talks about the unique attributes that different malts bring to the beer his preferences and particular style are slowly revealed. "Debittered dark malts have interesting properties that can be explored." He feels that by using a small amount of these malts in lighter beer styles both subtle flavors and incremental pH gains can be realized. Where they might not have been in large quantities, to Bill these specialty malts are the spice of the mash. For many years, a hallmark of his particular brewing style would include the use of a bit of wheat or oatmeal. "I felt that the protein added something to the body. Perhaps it was just my homebrewing roots but that is what I did." He is a particular fan of the flavor that kilned wheat malts bring to his beers.

Bill likes to plant a small plot of ornamental barley at his home every year. As it slowly grows it "has a calming effect" and provides a connection and allows a better understanding some of the work that goes into getting this vital material from the farm to the brewery.

Conclusion

Extract and color are but two of the many parameters that can be analyzed, calculated, managed, and optimized when brewing a beer. Malt flavor is tougher to get a handle on and is not quantifiable in the same way. The subtle interactions of malt flavor ultimately give the beer its harmonious beauty. It is often said that brewing is a blend of art and science. This is especially the case with malt and how brewers use it. Understanding the analytical contributions from malt is needed to make beer consistently, but flavor is ultimately what my goal is and that needs to come from personally experiencing the ingredient as it expresses in beer.

3

History of Malting

"I confess it facile to make barley water, an invention which found out itself, with little more than the bare joining the ingredients together. But to make malt for drink, was a master piece indeed."

—Thomas Fuller[1]

Humans have been malting grains for thousands of years, whether intentionally or accidentally. Modern malsters have tried to manipulate, enhance, and refine the process, but ultimately malting relies on the natural biological mechanisms present in viable barley kernels. Malting techniques have evolved from drying kernels on rocks heated by a fire to kilning in state-of-the-art machinery. But regardless of the time period or technology used, the goal has remained remarkably static: to transform a nearly indigestible kernel of grain into an ingredient for making beer.

Ancient History

Historians widely believe that the relationship between humans and grain predates recorded history, and a primary cause for the movement away from hunting and gathering as well as the development of settlements came from the desire to farm cereal grains. Cereal grains provided a steady and reliable source of food, allowing early humans more social and biological stability. While properly dried raw grains store well because of their dense, tough nature, they need some type of processing to make them more suitable for human consumption. To capitalize on the nutrients locked away in raw grain, humans experimented with preparation techniques to make grain easier to eat. All of the processes involved heat and water. Three distinct steps are required to make breads: grinding, mixing with water, and baking. Similarly, gruel or porridge is made by grinding grain and cooking it in water. Although humans realized that grains may be parched by placing them on a fire-heated rock to make them easier to grind, another, simpler method of preparing grains that did not require a cooking step was also discovered. If the grains were soaked and allowed to sprout they softened and became more palatable. It is likely that this was how the first stored grains were consumed. It is also conceivable that wild yeast and bacteria colonizing these gruels produced the very first beer.

Exactly where and when this occurred is not known, but archaeological evidence indicates that humans were gathering and consuming both barley and emmer (an ancestor of wheat) at least 23,000 years ago.[2] The Natufian cultures of the eastern Mediterranean predate the development of agriculture, but evidence indicates that in addition to the domestication of dogs,[3] these semi-sedentary hunter gatherers also developed all the technology needed to brew beer 12,000 to 15,000 years ago.[4]

As organized civilizations developed, so too did brewing techniques. Beers made with methods derived from Egyptian and Sumerian records have been re-created by modern researchers with acceptable results. Despite advances in brewing techniques and technology, one of the main challenges ancient brewers had to overcome was converting stored carbohydrates into fermentable sugars. These brewers realized that raw or dried grain did not produce beer unless further modified, and began experimenting with methods to extract sugars from the kernels. Roughly 3,800 years ago in Mesopotamia, the "Hymn to Ninkasi" was recorded on a cuneiform tablet, showing how steeping, germinating, and baking were used to produce "bappir;" sweetened barley bread that was mixed with water (mashed) and acted as the base of their beer.[5]

These techniques are still in use in the Nile area today. The English word "booze" is derived from "Bouza," a beer made from bread and malt in the Nile valley. Although most Western malts undergo a kilning step to dry them after germination, these primitive malting processes often finish with the germinated grain being used green (undried) or with the kernels drying in the sun.[6]

Accounts from the Roman Empire indicate that beers, and thus malts, were being made in Northern Europe. "A record of brewing in the fifth century says that the grain was then steeped in water, made to germinate, and was afterwards dried and ground; after which it was infused in a certain quantity of water, and then fermented, when it became a pleasant, warming, strengthening, and intoxicating liquor; and that it was commonly made from barley, though sometimes from wheat, oats, or millet." [7] It is clear from this description that although brewing has undergone many developments since, the basics of malting were already well practiced by the early Middle Ages.

Early Malting

Most malting during the Middle Ages was done on a small scale. Malting and brewing were domestic chores performed mainly by the women of the household, with the skills being passed down from mother to daughter. William Harrison's "Description of England" includes an extensive account on the making of malt:

> *"The best barley, which is steeped in a cistern, in greater or less quantity, by the space of three days and three nights, until it be thoroughly soaked. This being done, the water is drained from it by little and little, till it be quite gone. Afterward they take it out, and, laying it upon the clean floor on a round heap, it resteth so until it be ready to shoot at the root end, which maltsters call combing. When it beginneth therefore to shoot in this manner, they say it is come, and then forthwith they spread it abroad, first thick, and afterwards thinner and thinner upon the said floor (as it combeth), and there it lieth"* (Harrison, 1577)

After a minimum of 21 days, the germinating grain was finally ready for the kiln.[*] Malting was an essential skill across the English countryside. The 1623 publication "Countrey Contentments, or the English Huswife" by Gervase Markham, devotes 27 pages to the construction and operation of malthouses.

---

[*] Although four or five day germination time is the standard for malts today, until the industrialization of malting both the barley and the process used were different, and long germination periods were the norm.

In contrast, the section on brewing is a scant four pages long.

The earliest account of English malting can be found in a thirteenth century poem, called the "Treatise of Walter de Biblesworth":

> *"Then steep your barley in a vat,*
> *Large and broad, take care of that;*
> *When you shall have steeped your grain,*
> *And the water let out-drain,*
> *Take it to an upper floor,*
> *If you've swept it clean before,*
> *There couch,' and let your barley dwell,*
> *Till it germinates full well.*
> *Malt now you shall call the grain,*
> *Corn it ne'er shall be again.*
> *Stir the malt then with your hand,*
> *In heaps or rows now let it stand;*
> *On a tray then you shall take it,*
> *To a kiln to dry and bake it."*[8]

Regulations, court documents, and other official records give insight into reasons for and concerns about early malting. For example, malt kilns presented a constant threat of fire, so it was mandated that tubs of water be kept at the ready to douse any malt-born flames.

Regulation also began to dictate the quality of grain used, to ensure clean, drinkable beer. In 1482 the City of London ordained that malt must be "clene, swete, drye, and wele made, and not capped in the Sakkes nor Rawdried malte, dank or wete malte, or made of mowe brent barly, belyed malte, Edgrove malte, acrespired malte, wyvell eten malt or medled."[9] "Capped in the Sakkes" referred to the practice of deceitfully hiding goods of inferior quality at the bottom of the sack under better material. The translations of the other qualities include clean, sweet, dry and well made, not raw dried, wet, made of unripe barley, swollen, overgrown, or weevil eaten. Insect control was a frequent concern for early malsters; one account from 1577 notes that if malt was not properly "dried down, but slackly handled, it will breed a kind of worm called a weevil." [10]

The quality of malt that was produced in these times was often poor. The Court of Nottingham recorded a lawsuit alleging that on August 9th of 1432,

Thomas Sharp sold malt "raw reeked and damaged with weasles" (kilned raw and infested with weevils) to Thomas Abbot. The resultant beer "could not be held nor digested by them" and "hogs, hens, capons were therewith killed."[11]

Early Modern Period

Tyron's 1690 book, "A New Art of Brewing Beer," is the earliest manuscript specifically focused on brewing technique in the English language. The classic tome "The London and Country Brewer"* is widely available online and is a highly recommended read for brewers interested in the history of the craft. Despite their age, both books contain detailed instructions for malting processes that are still being used today. Modern brewers may find some of the information in the books surprising, especially the lengths and painstaking labors of some parts of the process. Tyron noted that grains were steeped for three full days and germination could potentially last for twenty-one days. The other book recommends that the malt on the germination floor be turned twelve to sixteen times per day by barefooted laborers to limit the damage to the malt.

Fuel types used to kiln the malt were also covered in these early works; the relative merits of coak (coke), Welch-coal, straw, wood, and fern are discussed, comparing and contrasting the benefits and drawbacks of each to the drying process. Although authors differ in their preferences, the inexpensive fern was generally panned for imparting a "rank disagreeable taste."[12] Ellis provides four simple methods for malt quality assessment: friability, steely ends (under modification), aker-spire (acrospire) length, and density, all of which can all be quickly assessed using just a bowl of water and a pair of teeth.

These early works give a palpable sense of a time when and place where the natural sciences were being tested and investigated. Experiments with elevated temperature drying showed that "when the fire on the kiln is excited with more vehemence, and kept up a longer time, it affects both the Salts and the Oils of the grain, in proportion to the degree of heat, and to the time, and thus occasions it to differ in colour; for fire, (says Sir Isaac Newton) and that more subtle dissolvent, putrefaction, by dividing the particles of substances, turn them black."[13] While this hypothesis was

---

* Although it was published anonymously, most scholars agree that it is very likely that William Ellis was the author of the work. Among other indications, the author gives a ringing endorsement of a pair of books about farming by William Ellis.

correct, many years would need to pass before scientists, such as French chemist Louis Camille Maillard, would definitively prove that color was the result of the interaction between amino acids and sugars.

Fig. 3.1: Detail from Hendrik Meijer of an early malthouse scene. ©Hendrik Meijer, DeMouterij Museum De Lakenhal Leiden

Early 19th Century

By the 1820s the English brewing industry had undergone major changes. Brewers began to apply more and more scientific method and inquiry, and the first published works describing both the use of hydrometers and elemental chemistry in brewing appeared on brewer's bookshelves. Not all of the at-the-time scientific findings would stand the test of time, as evidenced by George Wigney's assertions about oxygen, "Oxygen is the principle of acidity. . .united with beer, (oxygen) is productive of acetous acid (vinegar)."[14]

But even with good intentions, scientific progress is seldom easy, and not always readily embraced. When the young James Baverstock brought the thermometer into his father's Alton brewery, "he was forced to conceal and to use [it] by stealth to avoid parental outbursts about 'experimental innovations.'" In 1768, Baverstock acquired the recently developed hydrometer, despite his father's protests, and found the instrument very effective. He met with Mr. Samuel Whitbread, the principal brewer in London, to share his research on hydrometer use in the brewing process. Whitbread, uninterested in a novelty, dismissed him with a curt admonishment, "go home, young man, attend to your business and do not engage in such visionary pursuits."[15] Fortunately for all modern brewers, Baverstock did not take Whitbread's advice, and by 1824, the hydrometer was so widely used and accepted as a brewing tool that the government used it for taxation purposes, basing taxes upon ranges of wort strength.

The transition from home-based agrarian malting to larger scale commercial malting did not keep pace with the growth of the commercial brewing industry. The thermometer was largely accepted inside the brewery, but maltsters were slow to accept it. Wigney noted his frustration with the luddite nature of some malthouses, "I cannot quit this subject without strongly recommending a general introduction of the thermometer into the Malthouse...By this simple and easy acquisition, the art of Malting would no longer be governed by the errors and prejudices of illiterate and ignorant operators."[16] Because of their unwillingness or inability to embrace new technology and processes, many malting operations were not held in the highest regard by authors of the day, as Baverstock noted, "Malting is an operation generally confided to an ignorant labourer, who steeps his barley and turns his floors mechanically, and without regard to any other rule, than a certain number of hours, for each operation."[17]

Practically all malt made up until the start of the 20th century was raked out by hand. Because of the lack of mechanical refrigeration, malt was generally only made between October and May, with a fresh piece starting every three to four days. A single man could operate a ten to fifteen "quarter"* operation which would yield roughly two tons of finished malt per batch. Prior to the industrial age, where and when malting was done depended largely on the weather and the seasons. Malting was only a part-time seasonal job like hop-picking; farm laborers often acted as the workforce for the malthouses during the idle winter months.

Kilning technology developed differently in Germany, the United Kingdom, and the Americas. The British relied on direct heating from exhaust gases well into the 20th century. Indirect heated kilns were in use much earlier in Germany, with some systems being developed as early as the 1820s. The architecture and physical arrangement of drying kilns evolved over time. Thin layers of green malt spread on a horsehair cloth gradually gave way to perforated clay tiled floors. Perforated metal, wire cloth, and wedge wire floors replaced the tile floors during the second industrial revolution circa 1875.

The introduction of boats and trains allowed brewers to source malt from, and transport beer to, a greater area than before; in turn, production of ales from Burton increased nearly twenty fold between 1840 in 1870. At one point, Burton required grain from 100 local malthouses to support their levels of production.[18]

The British government, through taxes and regulation, helped shape the British malting industry. As early as 1325 the making of malt was taxed and regulated by the Crown. In 1548, King Edward VI decreed in *A True Bill for the Making of Malt* that the malting process must take at least 21 days. The malt tax—passed in 1697 and in effect until 1880—provided explicit instructions on how to malt, and generated considerable revenue. Taxes on malt and beer provided a full quarter of the total tax revenue (about double the contribution from land taxes), which helped fund colonial expansion and war efforts.[19]

Maltsters (especially those working in Scotland) were not pleased with the taxes and regulations. In 1724, the government sought to raise 20,000 pounds with a malt tax of sixpence a bushel in England and three pence a

---

* A "quarter" was originally a volume measurement. Eventually it became equivalent to a weight of 336 lbs. See also http://barclayperkins.blogspot.com/2010/09/weight-or-volume.html.

bushel in Scotland. Any taxes not collected to meet that amount became the burden of the maltsters.[20] Noncompliance with tax law was so great that in 1725 only 11 pounds and two shillings were collected. In Edinburgh, when the government sent soldiers to enforce payment, they were met by a rioting mob, and ultimately nine lives were lost over the taxation of malt.[21]

Because of the large amount of revenue involved, substantial regulation and observation was put into place to assure compliance throughout the United Kingdom. The law required production and stock records, advance written notice of intention to malt, specified hours of production, and free access by tax agents. By 1827, the malting laws included 101 individual penalties, each accompanied by hefty fines.

The needs of tax agents influenced the construction of malting facilities in the UK, which effectively halted innovation. At the time, malting operations involved steeping, couching, flooring, withering, and kilning. Steeping took place in a watertight cistern of specified dimensions usually made from iron, slate, or brick. After three days the moist barley was moved to the couch, an open top box where germination would begin and where the local tax agent would measure the depth of the wet and swollen grain to determine the duty to be paid by the maltster. After the first rootlets appeared, the chit malt would be moved to the germination floor. As germination finished, the grain was allowed to dry out or wither prior to kilning.

The tax agent would make measurements several times during the process. Physical volume was recorded for the tax calculations during steeping and couching. Because grain volume increases throughout hydration and growth, a wily maltster could lower their tax assessment by reserving water from steeping and couching, and sprinkle that water on the germination floor to finish the growth. As a result, the Crown enforced a rule stating that no water was allowed to be added to the grain for twelve days after steeping unless the steeping exceeded fifty hours, in which case the no-sprinkling window was adjusted to four days.

Any maltster caught attempting to compress or force grain in the couch would be fined the considerable sum of 100 pounds. If a malthouse worker broke any of the codified rules then they would be arrested, and upon conviction serve a three to twelve month sentence, "during the whole time for which he shall be committed, be kept to hard labour."[22] Historical accounts make it clear that the regulations were difficult to comply with and badly regarded by the industry.

Although the majority of historical documentation in the English language from this era concerns the British Isles, malt and beers were being produced worldwide. The brewing industry in the United States was quite small prior to the Civil War, and colonial citizens preferred hard cider to beer. In 1810 the total annual output of US breweries was less than 200,000 barrels. The industry had grown by 1850, but the country still produced fewer than one million barrels. Over the next 50 years, as the US population quadrupled, beer production swelled one hundred fold. Mass emigration of Germans to the US between 1830 and 1890 fueled this growth. Population pressures combined with the revolutions of 1848 (and eventual unification of Germany in 1871) influenced and inspired the development of robust brewing and malting industries in the upper Midwest.

Prior to this surge, brewing in colonial America had closely followed British techniques. One text instructed that in addition to guarding against rat infestation, the post revolutionary brewer should be aware that if "weevils at any time get into, or generate in your malt, which is common when held over beyond twelve or eighteen months, the simplest and easiest way of getting rid of them, is to place four or five lobsters on your heap of malt, the smell of which will soon compel the weevils to quit the malt, and take refuge on the walls, from which they can be swept with a broom into a sheet or table cloth laid on the malt, and so taken off. It is asserted, that by this simple contrivance not one weevil will remain in the heap."[23] One can only wonder if the flavors of insect infestation, rat excrement, and putrefying crustaceans contributed to the paltry beer consumption figures of those early American beer days!

Innovations of 1880

The 1880s heralded a period of tremendous economic expansion known as the Gilded Age. Rapidly developing technologies like electricity, railroads, and skyscrapers began to redefine modern life. The Brooklyn Bridge became a highly visible example of the power engineering and machinery had to fundamentally change humanity, and during this industrial era, mechanical power slowly replaced human muscle. Iron replaced wood, and malting evolved from a trade where a single person had deep understanding and ownership over every aspect of the process to an industry that was commercialized, profit-oriented, and able to be performed on a larger scale by many unskilled laborers. The long-standing English malt tax was repealed in 1880, which, combined with groundbreaking technical developments, would usher in a new era of malting.

As cities grew, the breweries and malthouses had to expand to meet increasing demand from larger urban populations. Malt at large was still made in the traditional manner: spread out on floors and moved by hand. The temperature in the malthouse was maintained by opening and shutting windows or louvers, by breaking up the couch, and by plowing and turning the piece. Because of the number of people and hours of labor involved, "the success of a malthouse [depended] in no small degree on the foreman in charge…and he [was] answerable for all the men employed in his particular house."[24] Unable to meet the surge in demand, the disadvantages and time investments of manual labor based malting became much more evident.

A Time for Malting

In 1934, shortly after Prohibition was lifted in the U.S., Arnold Wahl detailed some of the problems that floor malting operations experienced.

> "The quality of the malt as an end product was too much dependent on factors over which the maltster had no control like those occasioned by climate or temperature and weather conditions so that successful malting in the temperate zone was really only possible during the spring and fall months. Thus only during five or six months were the conditions favorable for malting. The control or regulation of the operations was uncertain. During night time malting the laborers often neglected their duties to the detriment of the quality of the product. Thus also upon the advance of spring incessant turning of the couches became so irksome the maltster's laborer was apt to leave his job to find another more congenial one in the brewery where more men were needed at that time. What could the malting establishment administration do in the calamitous situation when 50 to 100 tons of materials were contained in the steep, on the floors and on the kiln with the laborers leaving them when they were most needed? In cases of strikes in which the men were obliged as union members to obey the mandate and quit their jobs, the administration found itself hopelessly foundered."[25]

Many of the creatively engineered and elegant solutions devised to deal with the limitations of floor malting (namely the requirements for large amounts of manual labor and physical space) are still in use today. Three Frenchmen—R. d'Heureuse, Nicholas Galland, and Jules Saladin—each developed seminal pneumatic malting technology. The patented d' Heureuse air treatment allowed free circulation of air for barley germination. Galland made the critical connection that by using a regulated supply of water-saturated cold air, barley germination could be managed and the generated CO_2 carried away, and his patent for the process was granted in 1874. Galland next turned his attention to the problem of turning the malt; the first of his drum malting systems was installed in Berlin in 1885, and by 1889, one had made it all the way to Milwaukee. Galland-Henning drums were large steel cylinders resting on rollers, equipped with forced air ducting that turned periodically to homogenize the malt and break up any matted root material. Malting plants hosted banks of drums; the amount of malt varied from malthouse to malthouse, but barley charges of 10,000 pounds were not uncommon. Despite much success at the turn of the 20th century and further refinements by malting equipment inventors Tilden and Boby,[*] drum malting systems are rarely seen in modern malting.

Galland's research and developments dramatically improved the quality of malt (and logically the quality of the beer), so much so that by 1882 the German author and malt specialist Thausing noted that:

> *The quality of the malt is, according to all accounts, extraordinarily improved, as the entire process of germination can be carried through at such a uniformly low temperature as may be desired. In such malt-houses as Perry's, where malt-houses of the old and new systems are used at the same time, the difference in the smell of the germinating malt is very striking, the mouldy smell always perceptible in malt manufactured in the usual manner is entirely absent in that produced according to Galland's system.*[26]

---

[*] Your teenage kids will think that you are some kind of über malt nerd if you audibly express excitement at the sighting of the "Robert Boby Way Car Park." For those looking to horrify their own children; it is located in the historic malting town of Bury St. Edmunds in the U.K. and is near the Greene King Brewery.

Of the three inventors, Saladin contributed the longest lasting innovations. His compartment-based malting technology is still widely used today. "Saladin Boxes" are rectangular, open-topped germination compartments. An air plenum rests below the perforated false bottom that the grain rests on. Humidified, cooled air is forced through the grain bed, which can exceed 55–60 inches in depth. This increased depth meant much more malt could be dried at a time, as four inch-deep floor malting required considerably more area for the same total volume.

Saladin developed his solution to the problem of turning malt as he absentmindedly turned a corkscrew in a salt container during dinner one evening. In addition to the airflow, Saladin boxes are equipped with large rotating helical screws mounted on a carriage that slowly traverses the germinating grain bed. Saladin's design, with refinements from Prinz, an engineer based in Chicago, became the "Saladin Prinz" system and was widely adopted in the US. A keen-eyed historian can still find the nameplates from Saladin Prinz systems decorating the side of these marvelous machines in older malthouses all over the country.

Fig. 3.2: Saladin-Prinz nameplate in a shuttered malthouse.

Fig. 3.3: An early conceptual drawing of Saladin's original design. (Stopes, 1885)

Across the English Channel, Henry Stopes (1854–1902) contributed substantially to malt and brewery engineering. His insightful 1885 publication *Malt and Malting* was a technical tour de force of the rapidly changing industry and remains one of the seminal works on the subject. Henry was a man of prolific energy. He was an architect, paleontologist, brewer, and the father of Marie Stopes, a famous British advocate of birth control.

After spending his honeymoon touring continental European breweries, Stopes noted that "no greater mistake can be made in a brewery than to have it inadequate to the required work"[27] and set about designing and building a brewery in Colchester, Great Britain. During the 1880s, he designed or modified many malting plants using his own systems. Stopes regularly lectured on the subject of brewing and malt. Using modern architecture techniques, Stopes designed the technically advanced Barrett's Vauxhall Brewery which even featured an illuminated, rotating beer bottle atop the 119 foot brewhouse tower. It was the tallest brewery in Europe at the time, and featured a 147 foot chimney that was fashioned to look like his patented screw stopper. While other designers thought his architecture lacked sophistication, no one could argue that the four quarter brewery's entirely gravitation based system (that included no pipe fittings of any kind) was nothing short of an engineering marvel.

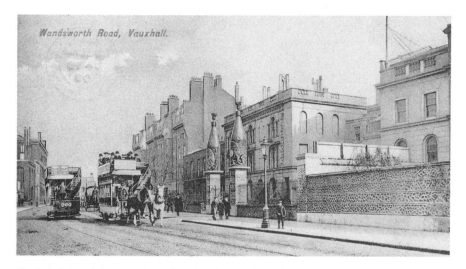

Fig. 3.4: Barrett's Brewery & Bottling Co. 87 Wandsworth, Vauxhall, England. Using the latest equipment of the time period, Henry Stopes designed part of Barrett's Brewery and included rotating beer bottles at the entrance. Courtesy of Richard Greatorex.

Taxes and Regulation

The introduction of the "Free Mash Tun" movement (which shifted the taxation point from the germination couch to the brewhouse) provided an economic incentive for the use of poor quality barley and malt. British grain farmers had foreseen the tax law change and incorrectly anticipated it would result in increased sales. Much to their chagrin, maltsters and brewers instead chose to purchase less expensive foreign grains. Barley from Turkey, California, Chile, India, and the European continent made its way into British malt and brewhouses. Brewers felt that the imported, bright colored six-row varieties improved the stability of the beer. This change also encouraged the use of adjuncts in English brewing, thus further depressing domestic barley markets.

The period's practical instructions for mashing malts included precautions against using "forced, slack, mouldy, steely (undergrown), overgrown, under-cured, or over-cured" or any other low quality type of grain. Maltser and author Thatcher offered some simple advice for brewers; "if compelled to employ mouldy grain, the brewer should endeavour to manipulate the material in such a manner that the flavour of the resulting beer is not spoilt. I advise the brewer to use the following useful blend for this purpose: 50 percent mouldy barley malt, 20 percent Smyrna, or some other well-grown, sound foreign barley malt, 10 percent flaked maize, and 20 percent sugar."[28]

Economically savvy brewers sought to make more consistent beer on a larger scale. To do so they needed consistent, competitively priced malt to be supplied at larger volumes. For small maltsters the squeeze was on; industrial scale malt operations were better able to meet the needs of the large brewer. Well-funded operations with ready access to the logistics needed to utilize less-expensive, foreign-grown barley had the competitive advantage.

Prior to "Free Mash Tun," country estates were exempt from malt taxes. The repeal of the malt tax negated the significant economic advantage for country estates to both malt and brew, leading to a decline in small-scale country brewing, and an increase in large scale industrial brewing.

This phenomenon was not unique to England or continental Europe; in the United States, taxes and regulations guided and shaped the industry. A portion of the barley needed to supply the rapidly growing US brewing industry was grown in Canada. In the 1882 *Report of the Tariff Commission* to Congress it was noted that while a 20 percent import duty was charged for finished malt arriving in the United States, a bushel of raw barley was subject to a flat $0.15 tax. After factoring in barley costs and malting losses, Canadian maltsters had a clear price advantage over domestic operations. As a result, malt imports soared from 144,487 bushels in 1875 to 1.1 million bushels in 1881.

Obviously the greatest governmental (and in turn economic) influence on malt production was to come many years later with Prohibition. The mass banning of alcohol meant a shift away from beer, and there was significant consumer interest in malts and malt extracts, presumably for baking. Despite the uptick in sales for home use, malthouses struggled, and like breweries, many did not survive the long drought in demand for beer and its requisite ingredients.

Later Developments

The vigorous debates recorded in the technical journals showed how passionate professionals were about barley. *The Wahl Handybook* (an important American brewing text), reads like a sales pitch, extolling the virtues of six-row barleys when used for the production of American beers containing cereal adjuncts. Most brewers today have a strong preference for low protein in barley and malt. Ironically, Wahl expressed the opinion that the low protein, two-row barleys grown in Montana and California were too susceptible to cloudiness, a view that is in direct opposition to modern understanding of beer haze. Lintner, a very influential and prolific German brewing scientist, took the position that ideal protein content of barley should be about 10 percent. His viewpoint has survived the test of time far better than that of Wahl, who was an advocate for higher protein levels of American barleys (12 to 13 percent) The well published brewing scientists Haase and Windisch* also became involved in this public discourse regarding barley protein levels. In a test utilizing barley from two different states (Montana with a protein content of 9.23 percent, and Minnesota with 15.16 percent), it was reported that the beer made from the high protein barley showed less sensitivity to chilling, greater durability after pasteurization, and greater palate fullness and foam stability than the beer made from the barley with low protein content. Later in Chapter 8 we will look much deeper into the history and development of barley varieties.

Tower malting systems were developed in the 1960s. Large circular germination and kilning chambers allowed for increased levels of automation and sanitation. Today there are far fewer workers and facilities producing far more malt worldwide than 100 years ago. These productivity gains would likely have been celebrated by the workers of old, despite the overall loss of jobs. In those days, employees turning malt in an active kiln were sometimes nude, wearing only cloth bags on their feet for protection from the heat.

References

1. Thomas Fuller, *The History of the Worthies of England*. (London, UK: Nuttall and Hodgson, 1840).

2. DR Piperno, et al. "Processing of wild cereal grains in the Upper Paleolithic revealed by starch grain analysis", *Nature* 430 (2004): 670-673.

3. James Serpell, *The Domestic Dog: Its Evolution, Behaviour, and Interactions with People*, (Cambridge, U.K.: Cambridge University Press, 1995).

---

* Lintner and Windisch were German brewing scientists who each developed scales used to measure the enzymatic power of malt. Both Degrees Lintner and Windisch–Kolbach Units are still used today for malt analysis.

4. Brian Hayden, Neil Canuel, and Jennifer Shanse, "What Was Brewing in the Natufian? An Archaeological Assessment of Brewing Technology in the Epipaleolithic". *Journal of Archaeological Method and Theory.* 20 (1) 2013:102-150.

5. Solomon H. Katz and Fritz Maytag, "Brewing an Ancient Beer". *Archaeology.* 44 (4): (July/August) 1991: 22-33.

6. D. E. Briggs, *Malts and Malting*, 1st ed. (London: Blackie Academic and Professional,1998).

7. W. L. Tizard, *The Theory and Practice of Brewing Illustrated.* (London: Gilbert & Rivington, 1850).

8. John Bickerdyke, *The curiosities of ale & beer: an entertaining history.* (London: Field & Tuer, 1886).

9. Reginald R. Sharpe (editor), "Folios 181 - 192: Nov 1482 - *Calendar of letter-books of the city of London: L: Edward IV-Henry VII*", British History Online:1899, http://www.british-history .ac.uk/report.aspx?compid=33657.

10. William Harrison, *Description of Elizabethan England, 1577,* (Whitefish, MT: Kessinger Publishing, 2006).

11. Bernard Quaritch, *The Corporation of Nottingham, Records of the Borough of Nottingham: 1399-1485.* Published under the authority of the corporation of Nottingham. (London:UK, 1883).

12. William Ellis, *The London and Country Brewer*, The 3rd ed., (London: Printed for J. and J. Fox, 1737).

13. Michael Combrune, *An Essay on Brewing With a View of Establishing the Principles of the Art,* (London: Printed for R. and J. Dodsley, in Pall-Mall, 1758).

14. George Adolphus Wigney, *A Philosophical Treatise on Malting and Brewing.* (Brighton, England: Worthing Press, 1823).

15. James Baverstock and J. H. Baverstock, *Treatises on Brewing,* (London: Printed for G. & W.B. Whittaker, 1824).

16. George Adolphus Wigney, *A Philosophical Treatise on Malting and Brewing,* (Brighton, England: Worthing Press, 1823).

17. James Baverstock and J. H. Baverstock. 1824. *Treatises on Brewing*. London: Printed for G. & W.B. Whittaker.

18. Christine Clark, *The British Malting Industry Since 1830*, (London, U.K. Hambledon Press, 1978).

19. _____, *The British Malting Industry Since 1830*, (London, U.K. Hambledon Press, 1978).

20. John Covzin. *Radical Glasgow: A Skeletal Sketch of Glasgow's Radical Traditions*, (Glasgow: Voline Press, 2003).

21. George William Thomson Omond, *The Lord Advocates of Scotland*, (Edinburgh: Douglas, 1883).

22. William Ford, *A Practical Treatise on Malting and Brewing*. (London, U.K. Published by the Author, 1862).

23. Joseph Coppinger, *The American Practical Brewer and Tanner*, (New York: Van Winkle and Wiley, 1815).

24. Julian L. Baker, *The Brewing Industry*. (London: Methuen & Co., 1905).

25. Arnold Spencer Wahl, *Wahl Handybook*, (Chicago Wahl Institute, Inc., 1944).

26. Julius Thausing, Anton Schwartz and A.H. Bauer, *The Theory and Practice of the Preparation of Malt and the Fabrication of Beer*, (Philadelphia: H.C. Baird & Co., 1882).

27. Lynn Pearson, *British Breweries-An Architectural History*, (Hambledon Press. London, U.K., 1999).

28. Frank Thatcher, *Brewing and Malting Practically Considered*. Country Brewers' Gazette Ltd., (London, 1898).

Floor Malting in Great Britain

Spreading a thin layer of wet barley on a bare concrete floor is the oldest germination technique for commercial malt production still in use today. The grassroots tradition of floor malting evokes a simpler time. There is also a lot of shoveling involved.

The town of Warminster, just south of the historic city of Bath, has a history of malting dating back to 1554. Warminster Maltings, founded in 1855, is the only one remaining of the 36 malthouses that have called the town home. In fact, there are only a few others left in the entire country. Tucker's Maltings, located about 100 miles east of Warminster in Devon, was established in 1831 and has been supplying locally grown and processed malt to regional breweries for almost 180 years.

The traditional floor malting process is very similar to the modern malt process, with several important differences. The steeping, germination, and kilning steps are typically separated, using different vessels, as opposed to combination steeping and germination tanks.

The germination stage probably has the most differences between floor malting and modern forced-air malting. In floor malting, the steeped grain is spread to a depth of approximately six inches (15cm) on

Turning the grain on the malting floor at Warminster remains a manual job. The step and pull method has been practiced by young men of the village for hundreds of years.

the germination floor. The grain is hand turned and plowed to untangle rootlets, dissipate CO_2, and control temperature. The piece is inspected several times a day, and the turning and plowing frequency is varied according to need. The process is very flexible and hands-on; the aim is to minimize variation across the floor.

In a modern malting plant, the germination bed will often be deeper than three feet (one meter). Helical screws turn the grain and humidified air is blown through a perforated floor, and then through the grain. While it can be argued which process has more variation in temperature and humidity, overall, floor malting is a slower, less automated process.

At Warminster, operations take place in a compact two-story building. Steeping is started by loading nearly 10 tons of barley into a brick lined depression and then filling the "ditch" with water. Over the course of two to three days, the water is periodically drained and refilled until the barley has reached the desired moisture level of near 50 percent by weight. The grain is then moved to one of eight 2000 square-foot concrete germination floors where it will remain for about five days. During that time, the rootlets start to emerge, and if allowed to grow unchecked, would become an impenetrably tangled, living mass that would eventually smother itself. The maltsters must turn and aerate the grain with manual plows or mechanical turners to keep the piece relatively free-moving, to keep the growth rate homogenous, and to prevent mold and mildew growth. Temperature and humidity control are managed by a combination of sprinkling with water, adjusting germination bed depth, opening or closing windows, and in the hottest days of the summer, the use of air conditioning.

In addition to the ever-present shovels, the employees at both facilities have other tools at their disposal for working the malt on the floor. The electric Robinson Turner resembles an old-fashioned push-type rotary lawnmower or rototiller. As its paddles turn, the machine redistributes the grain across the germination floor. "Ploughs" or malt rakes are manually pulled behind a worker with a jerking motion to lift and separate the mass. Although the Robinson Turner is used daily, as the germination nears completion, the manual plowing frequency increases to four times daily.

When it is time to move the grain to the kiln, a Reddler Power Shovel helps ease the burden. The power shovel is basically a horizontal plough blade that is pulled using a cable and winch. The operator guides the

A collection of malting tools at Warminster Maltings.

tool by using a pair of handles. Although clearly less taxing than using wheelbarrows and shovels, the job of moving malt remains fundamentally physical in nature.

Before the installation of the power shovel at Tuckers in the 70s, it took 12 men with shovels and wheelbarrows to move the heavy, wet, and sticky green malt from germination floor to kiln. Richard Wheeler of Tuckers remembers:

> *"If everything was working right and nothing went wrong, it took two to two and a half hours. But if everything did not go right, roots tangled together and the elevator got choked up. And then, after the whole load goes up to the kiln floor, four men with forks go up and spread it out. Oh, it was hard shoveling, before the kiln spreader, distributing the malt on the kiln floor."*

For both Warminster and Tuckers, the kilns have been upgraded to a more modern version than was originally installed. Tuckers replaced the coal-fire heated perforated tile in the 60s with wedge wire flooring and used heating oil as a heat source until 1980. A 12- to 15-inch bed of green malt is loaded into the kiln, and over the course of 40 hours it is dried (becoming pale ale malt) by blowing heated air up through the porous floor.

The floor malting process is revered by many brewers because they feel that malts made on the floor tend to have a more complex flavor than more-modern techniques, possibly due to microflora that is retained in the floor between batches. The barley variety probably has a lot to do with it as well. Chris Garratt of Warminster comments:

> "A few years ago we had commissioned a study at the Canadian Brewing and Malting Barley Research Institute to prove barley variety (Maris Otter, for example) had a real influence on beer character and flavor. The first year the malts were obtained from various malting houses in the UK but it was discovered that these malt samples produced beers of very different flavors and aromas – so much so that it was difficult to separate out the influence of barley variety. The floor malting samples were noticeably different in flavor to the factory malts for the same variety. The next year all the malts were made by the Institute and the variety differences were clearly measurable. The first years work wasn't published but I am confident that our floor malting process has a character that influences beer flavor and aroma. The aroma that comes out of a sack of Warminster malt is very distinct. Our maltings dates from mid 19th Century, I'm sure the very nature of the building and the prolonged malting methods we continue to use today all have an influence on the malt character."

4

From Barley to Malt

John Jablovskis, a regular at the Bell's Eccentric Café, recalls the simple malt recipe his family brought with them when they emigrated from Latvia in the 1890s: "Wet some good barley and keep it warm and moist. After it grows feet, form it into a rough loaf and place in a warm oven until dry."[1] All of the elements required to produce a carbohydrate source suitable for making beer are represented in this simple, traditional recipe. However, color, fermentability, extract yield, and virtually every other quality attribute would undoubtedly vary from batch to batch. While making malt is not difficult (in some form or fashion it has been done on a small scale in the home for all of recorded human history) making consistent, high grade malt under tight specifications presents quite a challenge.

Although a great gulf exists between the kitchen of Jablovskis' grandmother and a modern malthouse capable of producing 1000 metric tons per day, the fundamental steps of steeping, germination, and kilning are the same. The main objective in the steeping process is to raise the moisture content of the viable barley seed to a level that allows sprout growth. During the germination process, the barley seed is allowed to grow under controlled conditions. The kilning process then reduces the moisture content, halts the growth process, and develops the characteristic colors and flavors of the malt.

Throughout history, many different techniques and processes have been used to make malt. In this chapter we will examine the fundamentals of making malt, without getting too granular about specific practices.

Making Malt—Steeping, Germination, Kilning

Malting consists of three relatively simple steps: steeping, germination, and kilning. This process transforms a plant into a brewing resource; a ready, natural source of nourishment for yeast. As we will see in Chapter 8 the kernel consists of a plant embryo, densely packed reserves, and a protective husk. As the kernel germinates, the internal components are modified, creating enzymes and degrading the protein structure. In the controlled environment of the malthouse, hydration and dehydration are used to initiate and terminate the germination phase.

There are numerous supporting steps and functions in a commercial malthouse that come before and after steeping, germinating, and kilning. Before the barley ever reaches the steep tank, it must be purchased, tested, transported, stored, graded, and cleaned, sometimes repeatedly. Post kilning, the malt needs to have any rootlets broken off and removed before the malt is stored and eventually packaged or transported to a brewery. In addition, the malting process requires regular sanitation, testing, and quality assurance checks. Supporting functions like wastewater management are necessary to the operation but rarely of interest to brewers.

Throughout the history of malthouse design and malting optimization, the goal has been to reduce variation, process times, malting losses, and operational costs while increasing finished product quality. While process advances and skill have improved malting in the modern age, the products and ideas that came with industrialization played a large role in changing the world of malting as well.

Pre-Steep Activities

Acceptable barley must be acquired before any malt can be made. A critical factor is viability; in order to make malt, the barley must grow. Other important criteria include whether the barley is free from disease such as *fusarium graminearum* (sometimes known as Gibberella zeae), head blight that produces deoxynivalenol (DON), pre-harvest sprout damage, or insect damage. The grain must also have the ability to break dormancy, acceptable protein levels, uniform kernel size, intact husks, and an absence of broken kernels.

Barley may be physically stored in a grain bin on the farm, moved to offsite storage facilities such as a local grain elevator, or delivered directly to the malthouse. Regardless of its origin, the first operational steps in the malthouse are cleaning and grading. Barley may contain field trash and small quantities of wheat or other agricultural crops. Cleaning machines at the malthouse remove grain awns and loose straw, broken kernels, foreign seeds, small stones, trash, metal bits, dust, and chaff. During cleaning, barley will also be graded (separated according to size) before finally being stored in the receiving grain bins ready for use.

Steeping

The main objectives of the steeping phase of malting are to further clean and hydrate the barley. Steeping is performed either in a dedicated steep tank or in multipurpose equipment such as a Steeping Germination Kilning Vessel (SGKV). When the barley arrives at the malthouse it typically has a moisture content of about 12 percent. The steeping phase will increase the moisture level to 43 to 48 percent. The barley swells as the moisture level increases and can enlarge up to 40 percent (by volume).

Fig. 4.1: This Circular germination bed is 92' (28 m) across and 63" (1.6 m) deep. It holds 400 metric tons of barley per batch; the harvest from about 200 acres of local farmland.

The water used for steeping needs to be clean and of good quality. In many operations this water is warmed or cooled to a specified temperature, as the maltster needs to keep tight control of temperature during the entire process. As the grain is mixed with water, some of the microflora that is naturally present on the surface of the grain becomes waterborne and is removed. Straw or other lightweight field debris that has remained after the initial cleaning can float to the surface of the vessel. These unwanted materials are removed by allowing the water to overflow the steep tank into specifically designed collectors.

Fig. 4.2: Loading the Saladin boxes with grain from the steep tanks. The barley is moved as a slurry and its distribution in the bed is manually controlled at this malthouse.

As the barley soaks in the steep water, dirt, microbes, and other free materials make their way into the water, eventually tinting it brown. By draining the water two or three times during the steeping phase these contaminants are greatly reduced. Over the length of the soak, the barley kernels continue to absorb water, and their metabolism increases. Oxygen is needed to support respiration; if the plant embryo does not get sufficient oxygen it will drown and die. Air rests between active water steeping steps are used to provide the embryo with an opportunity to access oxygen. Air is often pulled down through the malt bed with ventilation piping during this aeration step, and helps carry away the carbon dioxide that is generated by the respiring barley. Compressed air may also be directed into the bottom of the steep tank to mix the solution, help float any leftover trash to the surface, and assure that sufficient oxygen is available to the barley.

An example of a typical 40-hour, three-water change steeping schedule in a modern malthouse is: 9-hour first immersion, 9-hour air rest, 6-hour immersion, 6-hour air rest, 5-hour immersion, and 5-hour rest. A cycle of this duration allows a fresh batch to be started every two days with ample time for cleaning. Shorter cycle times for the steeping process are often used during the summer months when warmer temperatures increase the metabolic activity in the barley.

Fig. 4.3: Steep tanks at a modern malthouse. 12 tanks are used for each 200 metric ton (441,000 lb) batch of malt. Saladin box germination beds are located directly below the steep tanks.

By comparison, the germination process one hundred years ago would take a couple of weeks, as noted in Chapter 2. Higher malthouse throughput is obviously more economical and has been a main goal in malting process enhancement. A greater understanding of the underlying science of malting coupled with process optimization and modern barley variety development has led to faster malting cycles. Despite the modern advancements (and malt's importance to finished beer), not all brewers feel that the quest for efficiency should have such a large focus in malting, and many now hunt for flavors that enhance the specific beer styles they brew.

Germination

The now fully-hydrated and activated barley is ready for germination. If the barley needs to move to a dedicated germination area, it is usually done after the water is drained away, but it can also be moved as wet slurry. Historically, all malt was germinated as a thin (3–6 inch) layer placed directly on a floor. In these systems, there is ample access to oxygen and the diffusion of carbon dioxide doesn't present much of a problem, as the grain bed is turned regularly. Germination bed temperature can be regulated by heaping up or spreading out the piece of malt. Modern pneumatic malt plants are much more space efficient; the germination bed can be up to 55–60 inches deep. All necessary ventilation is supplied by powered electric fans.

As the barley begins to grow, the tiny rootlets—or "chits"—emerge from the end where the kernel was attached to the plant. This sprout is the first visible sign of germination, and if the barley has had good aeration during steeping phase it may arrive at the germination bed already chitted. If allowed to grow unchecked, these rootlets will eventually tangle into a matted mess that hampers airflow through the bed, and eventually suffocates and kills the grain. To prevent grain death (and the rot that would soon follow) the sprouting grain needs to be periodically turned and the rootlets separated. Regular shoveling to turn over the germinating grain that typified floor malting throughout history has been supplanted by turning machines (prototyped by Saladin) in automated plants that lift, separate, and mix the developing malt.

Germination in modern pneumatic malting plants may be done in a germination drum, germination box, Steeping Germination Kilning Vessel (SGKV), or Germination Kilning Vessel (GKV) system. Malting drums are large cylinders that can be mechanically rotated to separate and mix the germinating barley. Invented in the late 1800s, malting drums have largely fallen out of favor despite their widespread use in early malting. More popular malting boxes are open-topped constructions outfitted with perforated false bottoms. Below the box is an air plenum which channels and directs the substantial quantity of air required. Rails mounted on the top of the box support a moving carriage equipped with rotating vertical screws. As the screw carriage slowly makes its way across the piece, the grain is gently broken up and turned over by the screws, preventing matting and hot spots. SGKV and GKV (also called fleximalt) systems combine several operations in a single vessel. The sequencing and timing of the various functions with these systems can be more easily varied and adjusted , meaning less movement from vessel to vessel (and less potential damage) to the grain. In these hybrid systems, the under-floor plenum is used for both the ventilation and kilning airflow.

Fig. 4.4: Production Director Dave Watson samples barley from the germination drum at French & Jupps in Ware, Hertfordshire, England.

Germinating vessels can be either rectangular or circular. Some newer malting plants are built as gravity fed, multi-floor "tower" operations, with stacked circular germinating vessels. When the grain is ready, it moves via a central shaft to circular germination compartments located on lower levels. The turning equipment is also used to load, level, and remove the barley from the compartments before it finally drops to kilns located at the lowest levels.

Water and air management are crucial during the germination phase. The air blown across and into the germinating kernels needs to be fully saturated with water to prevent the grain from losing moisture. Airflow is also used to

Fig. 4.5: Quality Control Manager Chris Trumpess with the impressive malting drums at the Simpsons malthouse in Tivetshall, St. Margaret. They are 12' (3.7m) in diameter and 48' (14.6m) long and hold 28 metric tons (61,730 lbs.) of barley each.

control the temperature of the grain bed. Heat energy is required to change water from a liquid to a vapor, and as the air becomes saturated with water the temperature drops. In some malthouses the incoming airflow moves through a chamber equipped with water nozzles. If necessary to maintain temperature, chilled water may be used to further cool the supplied air. In some situations, it becomes necessary to "sprinkle" the germination bed with additional water to help replace the moisture lost naturally by the respiring grain.

The predictable result of high humidity and abundant organic material is aggressive mold growth. Malthouse sanitation is a never-ending task; high pressure hoses and light bleach solutions become ubiquitous, and expected. The ample moisture also contributes to building and equipment degradation. Regular sanitation and careful selection of more corrosion resistant materials such as stainless steel are necessary to ensure equipment life and limit maintenance downtime.

As the barley germinates, fundamental biochemical changes occur within the kernel that affect both structure and composition. After being hydrated with water during steeping, the viable embryo begins to develop, and roots emerge from the proximal end of the kernel as the acrospire begins to grow between the endosperm and husk layer. If allowed to fully develop, the acrospire would emerge from the distal end of the kernel and form the main stalk of the barley plant. Functionally, the endosperm is the energy reserve that fuels the plant's growth, and consists of large starch structures held within a tough protein and complex carbohydrate matrix. As anyone who has attempted to chew raw barley can attest, the quiescent kernel is quite tough. As germination progresses, the protein material is broken down by enzymes generated by the husk layer. This process, known as modification, starts close to the embryo and eventually proceeds to the distal end of the grain.

A quick visual assessment of the elongation of the acrospire can indicate the degree of modification because the two factors are roughly in sync with each other. Another simple assessment of grain modification comes from "rubbing out" fully hydrated but ungerminated barley; modification breaks down the entire kernel's structure to a softer, dough-like texture. Simply rubbing a kernel between the fingers can show the extent of structural protein degradation. Enzyme activity and other starch and protein degrading substances will be explored in greater detail later in Chapter 6.

At the conclusion of the germination phase, the wet "green" malt needs to be dried to prevent mold growth or other kinds of spoilage. During the initial stages of drying, the rootlets "wither" as they lose moisture. In historic floor malting operations, this sometimes days-long stage of drying occurred on the germination floor, which from a modern perspective is considered as an additional step in the malting operational sequence. The term withering is used today to refer to the initial stage of kiln operation when the easily eliminated surface moisture is removed. In modern practice the bed is "stripped" (moved) to the kiln before moisture levels are allowed to drop significantly.

In contrast to floor malting operations, air moves through a pneumatic malthouse with the help of massive fans. Physical access to the various germination beds and kilns requires passage through sets of double doors that function as airlocks to hold in temperature. Pneumatic forces at play are quite amazing; it does not take much more than a few pounds per square inch of air pressure differential to make a standard sized door immovable by a person!

Gibberellic Acid

Many maltsters seem hesitant to discuss the use of Gibberellic Acid (GA). GA is a naturally occurring and very potent plant growth hormone that can trigger and/ or increase the speed of seed germination, and is commercially used in low dosages on a number of crops such as grapes. When applied early in the germination stage of malting it has dramatic effects and can allow efficient malting of otherwise problematic grain. So what is the big deal? Why not talk about it?

For many maltsters the use of GA is seen as a crutch, a confession that they can't manage the malting process. Dave Thomas, who headed up the malting operations at Coors for many years, provides a good perspective: "Like brewers, maltsters want to be seen as pure and natural. If they use GA, it's like an admission that they don't know how to malt the barley without it. It is a pride and ego thing. Gibb is a Band-Aid."

Bruno Vachon of Malterie Frontenac in Quebec knows that it is expensive, difficult to control, and prohibited by the Reinheitsgebot. He feels that a good maltster should be able to work with the raw materials that nature provides him. "It is like enzyme use in brewing; done by some but not frequently publicized." It is not surprising to find that many malt operations have a container hidden away for occasional use with a particularly problematic batch; it is a powerful tool in the maltster's arsenal. Usage rates are variable but 0.5 to 1 gram of GA per 10,000 pounds is a good ballpark figure.

On top of ethical conundrums and pride, GA is challenging to use; the application of tiny amounts to a massive pile of grain on a germination bed proves difficult, even for experienced malsters. The unchitted kernels that would benefit from GA the most are also the least able to uptake it. The result can be a highly variable, yet rapidly germinating piece of malt. It is easy to overuse which can result in overgown malt with high losses. Dave Thomas likens it to "throwing gasoline on a fire."

Some brewers will expressly allow or prohibit the use of GA on their malt. For the maltster faced with a lot of grain with high protein, low germination, or general water sensitivity, a little GA support can make a huge difference. Despite these advantages, if you asked a maltster the question in public they will likely respond, "We don't use GA."

Fig. 4.6: Saladin Box-type germination bed in the process of being "stripped" to the kiln.

Kilning

The main purpose of kilning is to remove moisture from the grain. By heating the kernels and removing the moisture, germination stops and colors and flavor begin to develop. The same variables that influence the other phases of malting are also used to manage and manipulate the kilning process, namely: time, temperature, and moisture. Using the additional control factors of airflow and the resulting degree of barley modification, a maltster is able to generate a wide variety of flavors in the malt via Maillard reactions (the result of amino acids reacting with sugars at elevated temperatures) and melanoidin development. During kilning some enzymes are destroyed, thus establishing the ultimate enzymatic potential of the malt. By controlling time, temperature and moisture, kilning practices strongly influence enzyme destruction. As an example, the cool temperatures used to kiln distilling malts are optimized for the maximum preservation of enzyme content.

Moisture must also be removed to store the malt for any period of time. As a bonus, the significant reduction of moisture content also reduces the weight of material to be transported, making post kilning logistics slightly easier.

Kilning operations can be roughly broken down into two phases: free drying (withering) and curing. The objective during the free drying phase is full-scale moisture removal. In short, the liquid water encapsulated in the grain kernel moves to the surface and evaporates. The evaporation process (a change of state from liquid to gaseous water) typically requires external heat. Warm, dry air is blown through the grain bed and is cooled as it picks up moisture. "Breakthrough" occurs when the majority of the moisture has been removed, the blown air is no longer being cooled by any moisture left in the grain, and curing begins. At this point, applied air temperature is increased and the malt first begins to develop color and aroma. If the temperature were raised while the barley was still wet, significant enzymatic destruction activity would occur, and very different malt would be produced.

Historically, malsters applied several different techniques to dry malt. Modern malsters have optimized the process; most malthouses today dry malt in kilns, by forcing heated air through a grain bed. Hot air has a greater capacity to carry moisture than cold air, and the amount of moisture present in the air can direct impact how quickly and efficiently the grain is dried. Quite simply, if the air enters the kiln with a high water content it cannot absorb much more. In winter, frigid, dry air contains very little humidity and when heated has a large capacity for moisture removal. Conversely, hot and humid summer air is not as efficient in removing water from the grain bed.

Despite modern advancements with climate controls, seasonal weather can have a significant effect on malt production. Malt made during hot and humid summer weather spends a proportionally greater time at higher kilning temperatures than malt made in the winter, ultimately resulting in malt with increased color. "Summer" malt could be made during the winter (by adding moisture to the dry air), but very pale malts (which require low humidity) are difficult to make in the summer.

The large variety of malts available to a brewer comes from a combination of specific cultivars and different drying techniques. Very pale malts are created using high airflow at low temperatures. Greater levels of modification combined with high temperature and moisture conditions results in darker malts. Initial drying at low temperatures followed by higher curing temperatures produce an entirely different set of flavors. The maltster can adjust many variables to produce a diverse range of flavors and colors in the kiln. Very pale malts may be kiln cured as low as 170°F, while darker and more aromatic malts may reach a final temperature of 230°F.

The use of combustion gases is the most efficient method to heat the kiln, and in turn the malt. Unfortunately there are negative effects of doing so; exhaust gases can contain several unwanted substances that can change, if not damage, the final product. Historical literature contains many accounts of brewers decrying unwanted flavors that came from the use of certain fuel sources. Although the transition to low sulfur coal in the early 1900s reduced the occurrence of these flavors, such coal often contained high levels of arsenic, which carried through into finished beer, and caused obvious, serious problems for brewers.

Off-Flavors

Oxides of nitrogen (NOx) are produced when combustion occurs in the presence of nitrogen (which is 80 percent of air). NOx will react with free amines in warm green malt to produce nitrosamines (NDMA), which are known carcinogens. Nitrosamines were finally identified as a problem in malt in the 1970s, and conversion away from direct heating with exhaust gases followed soon after. Since elevated NOx levels are found in the smog heavy ambient air in industrial and urban areas, nitrosamine mitigation and management becomes a constant challenge. The decision to close a large Los Angeles area malthouse was due in part to high NOx levels in the smoggy local air, which in turn caused high nitrosamine levels in the finished malt.

Sulfur, as SO_2 gas or derived from the combustion of elemental sulfur, is used in some kiln operations. It slows color development, lowers wort pH, and protects the enzymes during drying. It also minimizes nitrosamine production. Ironically, cleaner burning natural gas produces greater levels of nitrosamines than sulfur rich "dirty" fuels like oil. In contemporary North America, every large malt plant kiln is indirectly fired and no products of combustion come in contact with the malt to avoid off-flavors or the development of any dangerous substances. Indirectly heated kilns, the norm today, operate by transferring the heat from combustion gases to the air used for drying via large heat exchangers, meaning the air that contacts the malt is free of combustion products.

The characteristic cooked corn flavor of dimethyl sulfide (DMS) originates in malt. A small amount is an important flavor component of some beers (especially classic lagers), but DMS is quite volatile and easily lost in gas form from malt. DMS's precursor, S-methylmethionine (SMM) is neither flavor active nor volatile, but it does degrade at higher malt curing temperatures and as a result is not present at appreciable levels in hotter cured malts such as English Pale Ale or Munich. These "high dried" malts typically have lower enzymatic potential than their pale brethren.

Operation

The kilning operation begins with loading and leveling the green malt up to 35 inches deep on a perforated metal or wedge wire kiln deck. Heated air rising from the curing deck below moves upwards through the moist, green malt. As it does, the air becomes saturated with moisture and is vented outside the malthouse. The malt in the bed dries from the bottom up. When the piece has sweated out the easily accessible moisture, the free drying, evaporative phase ends, and breakthrough begins. Because energy is no longer being used to evaporate water, the temperature of the air exiting the malt begins to rise. During the free drying phase, ambient air may be mixed into the air leaving the curing piece below to adjust the temperature. Many kilns are operated on a two day cycle, using one day for free drying and a second for curing. At the end of the curing phase, ambient air is used to cool the piece for transport to cleaning and storage.

The floor of a double-decked, rectangular kiln is composed of sections roughly 30 inches across and as wide as the kiln, mounted on an axis. At scheduled times, a malthouse worker tips each section, and the malt falls to the level below.

Cleaning

After kiln drying, the thin, brittle culms (rootlets) are easily broken off using deculming equipment and then sold as animal feed. The malt is cleaned by sifting and screening before being placed in a storage bin. Freshly kilned malt often has poor performance in the brewhouse (most likely because of uneven moisture distribution throughout the batch), and is typically aged for at least three weeks. Milling and lautering are much easier when the moisture levels are allowed to equalize in a batch of grain over time.

The Result

Ultimately brewer preference dictates how the maltster guides the malting process. The vast majority of malt is made to be pale in color, with excellent enzymatic potential. By the time the malt has finished drying and rested, it contains all the readily accessible carbohydrates and Free Amino Nitrogen (FAN) needed to successfully make beer. The specific malt needs of individual brewers are heavily influenced by the type of adjuncts used (if any are used at all). There is a continuous spectrum of kilned malts that may be identified by similar names, meaning the properties and manufacturing processes of "pale" malt can differ considerably. It is hard to define where blue becomes green on the visual spectrum, and similarly there is no demarcation of where Pale Ale Malt becomes Munich Malt. Exploring and understanding the differences between malts is one of the great joys of being a brewer.

Malting Losses

There is a substantial loss of weight and mass as barley is transformed into malt. While it might seem like soaking and germination would *add* weight to the final product (with the absorbed additional water) 100 pounds of barley at the farm will become approximately 80 pounds of malt. The weight loss is explained in the following example:

- The original 100 pounds has a moisture content of about 12 percent, making the non-water weight 88 pounds
- Trash plus foreign and broken grains accounts for about 2 percent of the weight
- Respiration as the grain germinates accounts for 6 percent
- Rootlets (which are removed) account for 4 percent
- Final moisture content of malt is about 4 percent

As a result:

$$\text{weight as malt} = \frac{100 \times .88 \times .98 \times .94 \times .96}{.96} = \boxed{81.1}$$

Barley and malt may be bought and sold in bushel units. Confusingly, the original trading standards established by the United States Department of Agriculture (USDA) referenced weight by bushel (Bu), even though bushels are in reality a volume based measurement (specifically 1.244 cubic feet). The USDA trading standard for barley is 48 pounds per Bu and the trading standard for malt is 34 pounds per Bu. Thus 100 pounds of barley is equal to 2.08 Bu and malt typically produced from that amount of grain would have a finished weight of 81 lbs or 2.38 Bu. More simply put, 100 pounds of malt equals 2.94 Bu. This somewhat odd system is used only in the United States.

Moisture Management

For the hypothetical malt batch given above, the weight of the 100 pounds of incoming barley, less the moisture, is 88 pounds. If the malt is hydrated to 46 percent moisture, the total weight will reach 163 pounds. At 4 percent moisture (that of finished malt) it will weigh 92 pounds. The difference, 71 pounds, is the amount of moisture that must be removed during kilning. That's a lot of water to evaporate!

Conclusion

Although the simple steps of making malt consist of steeping, germination, and kilning, there are many facets of the process that can be manipulated to generate distinct flavors from the same base barley. Larger changes are also possible when additional processing steps are added, different raw materials are used, or other techniques are employed to make malt. Specialty malts (and their production) will be the subject of the next chapter.

References

1. Personal conversation at Bell's Eccentric Café, December 7, 2012.

5

Specialty Malts

The vast majority of base malt is made to supply the essential elements needed to produce beer; namely extract, FAN, and basic malty flavor. Specialty malts add diversity and complexity to a beer. Signature flavor and color are probably the two most noticeable attributes contributed by specialty malts, but these aspects only scratch the surface of the potential uses. Alternate grains, different processes, and additional ingredients can create a huge number of distinctive malts that change and enhance many aspects of a finished beer.

Specialty malt encompasses a diverse range of products. Broadly, the category covers any malt that is not a standard base malt, so specialty malt is in effect defined by what it is not. Many brewers break the larger category down into five subsets: high-dried, caramelized, roasted, alternate grains, and alternate processes. Although some brewing ingredients are labeled as adjuncts, they function more like specialty malts. Roasted barley is a prime example: since it does not go through the steeping, germinating, or kilning, it is technically not a malt, but the flavors and properties it brings to beer are quite similar to those imparted by black malt.

Flavor Development

The distinctive flavors of high-dried, caramelized, and roasted malts are created by changes in processing; the same source barley for these could easily be used to produce base malt instead. The characteristic flavors of specialty malts mainly come from Maillard reactions.

Much like making very dark toast, there is a fine line between success and smoke when making roasted malts. The old adage that making black malt is a good way to burn down a malthouse isn't that farfetched given that the temperature needed for their production is close to that of spontaneous combustion. As the maltster slowly raises the temperature, the amino acids and sugars in the grain react, causing the kernels to darken and develop new, complex flavors. When the desired color and flavor is finally reached, the process is quickly halted with a shot of cooling water. These sprays are also handy for quenching unintended combustion.

Roasting remains a hands-on operation and the timing of the endpoint is monitored by skilled operators who repeatedly remove samples for rapid color analysis. Color development starts out slowly, but builds quickly in the final stages. A single lot is generally composed of multiple batches, so the maltster usually has the opportunity to make subtle adjustments to meet a specification.

Some specialty malts require the use of specialized, dedicated equipment. Drum roasters are preferred for processing both coffee beans and malt. The rotating, heated cylinder of the drum is equipped with internal vanes; as it turns at about 30 rotations per minute, it thoroughly and uniformly mixes the material inside. Although small drums are used by local coffee roasting operations, much larger machines are generally used by commercial maltsters.

The drum roasters are also equipped with features that facilitate air and moisture management. A large burner assembly provides the heat source. High temperature exhaust gases are directed around the drum, heating the assembly indirectly. Additional fresh air inlets allow the maltster to pull fresh, cooler air around the rotating drum as needed. The ventilation system allows the maltster to control time, temperature, and moisture content variables inside the drum. These controls give the maltster more time to achieve the target color specifications without overshooting or burning the malt.

Heating green malt, fresh from the germination bed, causes the water in the grain to evaporate. Heating while constantly drawing fresh air through the chamber removes this suspended moisture, drying and toasting the malt. If the airflow dampers to the drum are closed, the moisture can be retained, which is a typical practice when making caramel malt.

Fig. 5.1: Repairing a bearing inside of a roasting drum. It can make up to 2 metric tons (4,410 lbs) of roasted malt per batch. Note the vanes used to mix the malt as the drum rotates.

When the malt is heated to a very high temperature, significant charring results in very dark colors and burnt flavors. Finished malt, not green malt, is generally used as the basis for black or chocolate malt.

Different kinds of high-dried malts such as pale, amber, and brown were well-established by 1736 when *"The London and Country Brewer"* was published. These malts were made using standard malt kilns, and differed in their production by variations in temperature. Brown, blown, or snap malts were made by quickly increasing kiln temperature, a process initiated by adding small bundles of wood to the fire.

The first equipment used to intentionally produce specialty malts was likely a simple pot or iron plate heated by an open wood fire. In 1817, Daniel Wheeler, a burnt sugar maker of London, patented "A New or Improved Method of Drying and Preparation of Malt" which was an iron roasting cylinder (likely the common ancestor of all modern roasting drums) that produced very dark malts. These "patent malts" created in Wheeler's cylinder performed better and were more efficient when brewing the popular porters, and were quickly adopted by cost-conscious brewers.

"If too quick or fierce a fire be employed to dry malt, instead of gently evaporating the watery parts of the corn, it torrefies the outward skin thereof, so raises the enclosed air as to burst the vessels of the grain, and divides the outward skin from the body of the corn, (such are called Blown Malts) by which means it occupies a larger space. If such a fire be continued, it even vitrifies some parts of the grain, from whence the malt is said to be glassy." (Combrune, 1758)

The evolution of malt roasting apparatuses has been directly influenced by similar equipment used to process coffee, chocolate, and nuts. Companies produce roasting equipment for these larger, more commercially viable markets, but maltsters and malt roasters benefit from any innovations. Malt specific equipment can vary greatly in size; batch sizes ranging from 300 to 5500 kg can be accommodated by modern roasting drums, depending on a maltster's need.

Over the years, roasting equipment has seen many variations on the same "heat and rotate" theme. The five "K Balls" used many years ago by Briess Malting & Ingredients Company are particularly memorable; these 1930s iron spheres rotated within a small chamber of flames. Each held about 800 pounds of malt and looked remarkably like wacky props from the set of a science fiction B-movie. According to maltster Dave Kuske, the first drum roasters went into service in the1970s, but K Balls continued to run for many years until they were finally retired in 2004.

Specialty malts can be produced on the homebrew or small brewery scale using an oven or smoker. An understanding of the driving factors behind malt flavor development (namely moisture, time, and temperature) provides all the information needed to experiment with and create different flavors. A word of caution: on a small scale spontaneous combustion at high temperatures is always a possibility. Always closely monitor roasting malts and keep a fire extinguisher handy.

Advanced Malt Flavor Chemistry

The products of Maillard reactions are integral to beer flavor, and the quantity and type of these products varies based on how the malt is kilned. The spectrum of compounds depends on the specific mixture of moisture,

Fig 5.2: K-Ball malt roaster in use at Briess Malting. Courtesy Briess Malt & Ingredients Company .

time, and temperature conditions, as well as green malt substrate. The task of the maltster is to make consistent products despite these natural variations. The temperature and humidity of the air needed to kiln or cool malt (as well as natural fluctuation in barley protein levels) force the maltster to adapt to maintain consistency or match brewer specifications.

Understanding the complex flavors and colors generated in malting requires a deeper understanding of the Maillard reactions which we will explore in Chapter 6. The compounds have flavor descriptors like acrid, bitter, burnt, onion, solvent, rancid, sweaty, cabbage, caramel, cracker, bready, bread crust-like, nutty, toffee, caramel, coffee, roasted, and malty.[1] The specific products are highly dependent on the concentration and types of sugars and amino acids, as well as temperature and pH. It may look as simple as making toast, but a lot of complex chemistry goes on inside the malt kiln.

High-Dried Malts

High-dried malts can be made in standard kilns by raising temperatures. Malt exposed to higher temperatures during the final stages of kilning will have a darker color and more malty/biscuit flavor than normally processed malt (Munich type malts are made this way). In addition to the time, temperature, and humidity levels, other factors like barley variety, degree of modification, and moisture level of green malt make it possible to create bread crust and biscuit flavors in floor dried malts.

Caramel Malts

To make caramel malts, the maltster begins with green malt, which is taken directly from the germination bed fully modified but unkilned. By raising the temperature of the still wet grain to enzyme conversion temperatures, the maltster effectively mashes the grain while still in the husk. The enzymes then continue the work begun in malting to break proteins into amino acids and starches into simple sugars. As the heat in the kiln is raised, Maillard and caramelization reactions begin, and a wide range of reaction compounds form.

Although it is possible to produce caramel malts on a kiln floor, there are significant flavor differences between the floor and drum versions. There are two main reasons for this; the kiln-based process is more uniform because the drum constantly turns the malt, and the temperature can be raised much more quickly in the drum than on the floor. Rapid water removal and quick rises in temperatures produce the glassy endosperm and clean, candy-like flavors that characterize crystal malts. Conversely, the slower water removal and temperature increases of the floor-based process tend to develop more malty/biscuit flavor notes.[2]

Historically, some caramel malts were made by placing a tarp over green malt to retain the moisture needed for conversion. Modern specialty kilns that have the capability to make caramel malts are fitted with air recyclers, and can reach 250°F. Standard kilns rarely have the capacity to exceed 190°F, and cannot produce malts darker than 60 Lovibond.

Roasted Malts

Roasted malts require a drum roaster because of the very high temperatures involved in the process. The drum can be loaded with green malt, finished malt, or even unmalted grains depending on the desired final product. Roasted malts exhibit flavors only developed at high temperatures

Fig. 5.3: When making caramel malts, an indication that the starches in the grain have been sufficiently converted to sugars is when you can "pop it like a zit."

Fig. 5.4: Tools of the Trade: well-worn hand grinder with samples used to visually assess roasted malt in process at French & Jupps.

(rich chocolate or coffee notes) but the process can be optimized to provide high color with minimum flavor. The spectrum of malts is very wide; the point where caramel malts end and roasted malts begins is somewhat arbitrary, but the transition occurs somewhere in the range of 325-350°F (160-175°C).[3]

Drum roasting equipment used to make roasted malts can also be used to produce caramel malts. The ability to quickly move large quantities of thermal energy into the processing grain allows the maltster to target and develop particular flavors. The physical process is rough on the grain; high heat and lots of tumbling can cause the grain to break apart, so appropriate barley variety and quality is critical to successful roasting.

The manufacturing of caramel and roasted malts remains an intensely hands-on job. The constant adjustment and assessment of these malts as they roast has not been automated and relies on an attentive, experienced craftsperson. Making specialty malts also requires significant investment of labor, capital, equipment, and time. These expenses (in combination with the relatively limited market for roasted malts) are one of the main reasons specialty malts cost more than base malts.

Making Specialty Malts

The processes used to produce specialty malts differ dramatically from those used to create pale malts. In the US there are five legacy malting companies; of these, only two have drum roasting capabilities for production of specialty malts: Great Western Malting Company in Vancouver, Washington, and Briess Malt & Ingredients Company, based in Chilton, Wisconsin. Cargill and MaltEurop have specialty kilns that allow for production of a few specialty malts. By malting company standards, Briess is a tiny operation. Their 45,000 metric ton annual capacity is dwarfed by the other four, but what they lack in quantity, they make up for in quality and diversity of their products.

Comparatively, Briess does not make much base malt, and a significant portion of their business is directed at food applications with products like specialty grain syrups. This focus gives them a unique understanding and comfort when working with difficult ingredients to produce unusual results. Dave Kuske heads up Briess malting operations. His wealth of practical malting knowledge comes from technical training and years of experience (he began working with malt in 1988).

Like any craftsman, Kuske credits much of his success to having great tools. Between the Chilton and Waterloo plants, Briess operates a total of five roasting drums. The weight capacity for these drums ranges from 5,200–10,800 pounds (2350–3350kg). Kuske knows that consistent green malt is a necessity to make consistent final malt and that barley variety and upstream malting practices are critical to quality specialty grains. "For caramel, I want a variety with a good solid hull. Once, I tried to roast Harrington and it was brutal." Kuske pays particular attention to hue. Beyond the simple Lovibond measurement (which just measures light absorbance), perceived color can offer a lot of information about the grain. "Hue is somewhat dependent on protein levels in the green malt. Higher protein barley gives a red hue; lower protein is more orange-amber."

The main reason that caramel malts have different flavors than roasted malts is the very high heating rates that drive caramelization reactions in the grain. According to Kuske, kiln-made caramel malts often include silage-like lactic off-notes. Although crystal and caramel are used to somewhat interchangeably when describing this type of malt, there are some notable differences. Caramel malts are a broader class that also includes kiln made versions. The glassy, crystallized interior characteristic of "true" crystal malts requires the use of a drum roaster. For Kuske, the differences between the two processes are largely defined by how quickly he can apply heat to a mass of malt, and how hot he can ultimately get it. Special malt kilns that produce caramel malts can reach maximum temperatures of 250–255°F (121–124°C), whereas drum roasters are capable of surpassing the point of spontaneous combustion for malt; about 460°F (238°C).

To make a batch of crystal malt, Kuske starts with fully modified green malt from the germination bed. The 68°F (20°C) malt is loaded into the roaster at approximately 40 percent moisture, and brought up to 158°F (70°C) over 15 minutes. Conversion completes after 40 to 45 minutes, at which point the drying phase begins. For about two hours, malt temperatures stay relatively low as water continues to evaporate. When the moisture level drops below 15 percent, color development begins, and by the time it reaches eight percent the color changes rapidly. Crystal malts reach an ultimate temperature of about 375°F (191°C). At this stage, the grain sugars condense into the crystallized structure that gives this malt its name. The difference between a 10°L and 120°L malt is only a matter of minutes in the roaster; the roast house operator relies on frequent sampling and visual assessment to determine when to stop the process.

Fig. 5.5: Chris Trumpess and Peter Simpson sample the still snapping and quite hot crystal malt as it is discharged from the roasting drum At Simpsons Malt in Tivetshall. The aroma was incredible.

When a finished batch is emptied from the roaster, the malt kernels audibly snap and a wonderful aroma immediately envelops the area.

The complex chemical changes that occur in the roaster also produce less desirable aroma compounds. After-burners are installed on the exhaust stack to re-combust gases at temperatures above 1300°F (700°C), and control volatile emissions. According to Kuske, these are needed "because above 300–320°F (150–160°C) you get yellow, acrid smoke, and unhappy neighbors." Operating an after-burner is expensive; one maltster estimated that for every two units of energy used to roast malt, another unit is used for the afterburner.

Black malt is created by roasting dry base malt. As it is heated to extreme conditions, many of the flavor and color reactions reach a point where the active components are either destroyed or driven off. Kuske notes, "On the way up to chocolate you get some nasty, bitter, acrid flavors and those get volatized and blown off. Why doesn't black malt taste like coffee? It's because you have volatized the flavors and lost the nice flavors. At some point you hit a wall and completely lose even the extractable color. If you are not careful, black malt can go from 600°L to 150°L in about a minute."

As Kuske makes malt he asks himself, "what predominant reactions am I looking for? Do I want sweetness, toffee, or viscosity?" Although flavor and color are linked, that linkage can be stretched or broken. Warm and wet conditions favor protease activity that results in darker colors. Malt made during a hot, humid summer has darker colors but no corresponding increase in malty flavor. "Summer Malt" is the term brewers who specify very lightly colored malts use for the darker malt they receive when it is made under these conditions. This malt demonstrates that when handled by a skilled maltster, an increase of color does not necessarily translate to additional malty flavor (which would obviously be problematic from a consistency standpoint).

Biscuit malts (such as the Victory malt from Briess) need to be used gently, because although they are loaded with nutty tasting pyrazine compounds, they have a color contribution of only about 30°L. Use at levels that impart significant color in beer will likely create an overwhelming flavor.

Ask Kuske what his favorite malt is and he'll quickly answer: "Briess Extra Special Malt. It is a hybrid between high roasted and caramel. I also like the flavors in Dark Chocolate." Both of these malts shine as examples of the full expression of the complex flavors available in the maltster's arsenal.

Other Grains

The malting process can also extract soluble sugars from the starches in other grains. Wheat is well known and widely used throughout the brewing industry, as is rye to a lesser extent. Most cereal crops can be malted, albeit with varying levels of ease and success. In barley, the enzymatic development and grain modification is fairly straightforward, and a well-made base malt has sufficient enzymatic activity to convert its own starches. In other grains, the amount of germination needed to make the starches accessible may result in significant loss of potential extract.

The very structure of barley (with its closely attached husk) makes it easy to process. Other grains, such as rye and sorghum, can be notoriously difficult to work with. It is difficult to achieve proper airflow through a heavy, closely packed, and viscous grain bed. If the grains cannot respire properly, they will die and rot, which (quite obviously) will negatively affect flavor.

Chris Garratt of Warminster Malting adjusts production schedules to be able to malt rye. He finds that as it hydrates, it releases gummy starches and "compacts tremendously in the steep tank." Unlike barley, moisture saturation of the grain sits below 30 percent and he has found that germination time on the floor can also vary anywhere from two to four days.

Lesser-known malted grains include oats, triticale, corn, rice, sorghum, and millet. Pulses and legumes like beans and peas, while not technically grains, can also be sprouted and malted.

Other Processes

Indirectly heated kilns are relatively new in the history of malting. Prior to the introduction of cleaner tasting fuels (such as coal or coke), kilned malts had a discernible smoky character. Although the flavor of wood fire has largely been eradicated, notable reminders of its legacy remain. When most beer drinkers think of a smoky beer, they think of Bamberg. Beers from Gotland, Sweden, Strasbourg, France, and the northern German styles of grätzer and lichtenhainer also have at least a hint of smoke.[4]

The type of fuel that creates the smoke malt has a large impact on specific elements of the flavor. Bamberg uses beechwood. The Alaskan Brewing Company brews a perennially medal-winning Smoked Porter, the malt for which is smoked over alder wood in facilities that used to process freshly caught salmon. Fruitwoods (such as apple or cherry) produce much sought after smoky flavors. In Scotland, malt made to produce the excellent whiskeys is smoked over peat. All of these malts tend to be intense, with high levels of phenolic compounds. They should be used sparingly in beer, if at all.

The German word saüre (the root of the English word sour) translates to "acid." Sauermalz (also known as sour malt or acidulated malt) is produced by intentionally allowing lactic acid bacteria to grow during the malting process, because changing brewing water pH by adding acid is not allowed under the Reinheitsgebot. These malts provide a means (that functions within the purity laws) to counteract the effect of alkaline waters.

Other Products
Dehusked/Debittered Malts
Husk material contains high levels of astringent tannins, which manifest during roasting. Production of very dark malts using dehusked barley reduces these unwanted flavors. Alternately, huskless grains like wheat can be used to make less astringent final malts.

Roasted Unmalted Grains
From a certain perspective, it seems wasteful to first go through the malting process, just to turn around and roast them to charred black, as black malt contributes no enzyme activity or fermentable extract to the mash. By the time roasting is finished, the organic material in the grain has been transformed into color components and the internal structure has been broken down to a highly friable state. Unmalted grains subjected to the same heat processing conditions function in a similar way from a physical standpoint. There are differences between black malt and black barley but these are much smaller than the gulf that separates unmalted barley from pale malt.

Pre-Gelatinized Adjuncts
The starches in the endosperm of the grain are tightly packed and must be expanded or made soluble before they are accessible to the enzymes. The structure can be opened using one of three methods: gelatinization, torrefaction, or flaking. Cooking rice or grits at home on the stovetop gelatinizes them. When a starch granule is heated in water, it swells and its structure is irreversibly altered. The temperature needed for gelatinization varies for different starches; for example, wheat gelatinizes at a lower temperature than corn.

Torrefaction modifies whole grain by exposing it to an intense heat source. During this process, moisture in the grain expands rapidly as it turns to steam, which causes an increase in volume and a reconfiguration of starch. Puffed rice breakfast cereals are an example of torrefaction; wheat, barley, oats, and even corn can be processed this way. Equipment for torrefaction and puffing operations generally consists of a conveyor traveling under an infrared heat source.

In contrast, pre-gelatinized adjunct flakes are produced by steaming the whole grain and then passing it through heated rollers. Both instant oatmeal and flaked breakfast cereals are produced using this technique. Like torrefied products, they can be directly added to the mash tun and excess enzymes from malt will convert the now accessible starches to sugar. These brewing adjuncts will lighten the flavor and color of beer and are much easier to use than brewers grits or rice (which require gelatinization in a cereal cooker before they can be used).

Malt Extracts

Highly concentrated syrups or dried malt extracts are produced by many maltsters (and even some brewers). In addition to fermentable malt extracts, highly colored syrups that add a certain hue to the final beer are available commercially. Although caramel colors are forbidden by the Reinheitsgebot, highly-colored, concentrated malt syrups like Weyermann's Sinamar are acceptable. Despite its huge color contribution (more than 3000°L) Sinamar has relatively little flavor, so it is a good product to be aware of when formulating a beer that requires a dark color but little or no dark malt flavor. [5]

Lagnappe (A little something extra)

Authors Note: Some of the more extreme malt flavors have excited the imaginations of brewers (and writers) throughout history. The following passage, from the perspective of a frequent brewer and drinker of stouts, is a particular favorite of mine. Prior to reading it I had no idea of the mortal danger I frequently place myself in.

"Now, as to the Complexion or colour of Mault, white is the best, because most natural, and therefore in all preparations and operations you ought, as near as possible you can, to maintain the

natural complexion of the thing, for the tinctures ariseth, and proceeds from the fine spirits and essential virtues; therefore, if in your order of making Mault, you alter and change the colour, you then also change its virtues, and make the Drink of another nature and operation; for all redness, or high Colour in Drink proceeds from some violence done to the Spirits and fine virtues in the preparation, for the colour is a stranger to the nature of that grain, and it shows that the fierce Spirits and hot vapours of the fire, have as it were transmuted or changed the mild friendly soft Virtues and Qualities of the Mault, into its own fiery nature; and it is not to be doubted, but that Dijestion is most natural that preserves the tincture entire, force not Nature out of her way, nor change the form, for then the inward Life and good qualities of that thing are in danger; for the fierce raging spirits of the fire, and Essences thereof do never depart from such parched high dried Mault, but do always remain, from whence the Drink made thereof receives its high bloody Colour, which most ignorant people cry up and admire as a virtue, or good quality, but the contrary is to be understood, and nothing in Mault is a greater Vice or Evil, and the Drink made thereof, together with its losing boiling with Hops, do seldom fail to wound the Health of the Drinkers thereof; its natural operation in the body, is to heat the Blood, destroying Appetite, obstructs the Stomach, sending gross dulling Fumes into the head, dulls the fine pure spirits, hinders the free Circulation of the Blood by Stagnating the Humours, and in the Cholerick and Melancholy Complexions, generates the Stone, Gravel, Gout and Consumption; this sort of Drink is also very injurious to women, especially such as are with Child, or give Suck. In all your preparations of Meats and Drinks use diligence, and do no violence to the internal Virtues or Spirits, which are occult and not perceptable to the Eye, therefore if you have not Wisdom and Understanding to distinguish the inward Virtues and Qualifications, and what properties governs in each thing ; you may as it were insensibly commit great Errors in the Smallest and Meanest Preparations, as indeed most do for want of an inward understanding of the forms of Nature ; and remember this that the nearer you come to Nature, and the more you imitate her, the nearer you are to

Truth. Now as to Barley, God and His Handmaid Nature have signed and indued it with that most Amiable Colour White, which of all Colours and Complexions have the first place ; therefore in your Preparation and Digestion of Mault use no violence that may cause it to degenerate, but on the contrary, use all Art and Means to maintain it, for the whiter your drink is the better and more healthful, having a mild or gentle Operation ; it is not so apt to send dulling, gross thick Fumes and Vapors into the head, nor to heat the Blood, or obstruct the passages ; and I must here tell you one truth ; namely, that White Ale of a middle strength drank new have the first place of all firmented Drinks, and is the best against the Generation of the Stone, Gravel, and Gout, excepting good, mild, soft Water, which is the radix of all moist nourishment."

-From A New Art of Brewing Beer
by Thomas Tryon published in 1691

References

1. Charlie Scandrett, "Maillard Reactions 101: Theory" http://www.brewery.org/brewery/library/Maillard_CS0497.html.

2. Terry Foster and Bob Hansen, "Is it Crystal or Caramel Malt?" *Brew Your Own*, Nov 2013.

3. S. Vandecan, et al., *Formation of Flavour, Color and Reducing Power During the Production Process of Dark Speciality Malts.* Journal of the American Society of Brewing Chemists. 69 (3), (St. Paul, MN: ASBC, 2011) 150-157.

4. Randy Mosher, *Tasting Beer: An Insider's Guide to the World's Greatest Drink.* (North Adams, MA: Storey, 2009).

5. J.C. Riese, "Colored Beer As Color and Flavor", *MBAA Technical Quarterly*, Vol. 34(2), (St. Paul, MN: MBAA,1997) 91-95.

Full Scale
Modern Malting

A distinctive aura surrounds older industrial malting installations. It isn't just the architectural style that gives each building its own distinct personality (although that is certainly part of it). Some were built to be grand marvels, others were created to be strictly utilitarian, while others yet seem to be overdue for a date with a wrecking ball. Even so, these facilities all seem to exude a wisdom gained from long years of use that shows itself in odd touches of grandeur and evidence of multiple instances of repurposing. The surprising ways that they can subtly reveal the depth and nuance of a long history will fascinate even veteran brewers.

MaltEurop
Milwaukee, Wisconsin

There is a malthouse located a few miles south of Milwaukee's modern and highly engineered baseball stadium. It seems massive compared to the small blue-collar homes that surround the facility, with towering cast concrete grain bins, constant railroad activity, and a pervasive low rumble from several massive blowers. Malthouse 1 was originally built

Barley and malt storage bins at Integrow Malt, Idaho Falls, ID. The metal bins on the right can hold 8,000 metric tons (17,600,000 lbs) of barley each.

in 1910; Malthouse 2 was built after the end of Prohibition. Malthouse 3 went up shortly after the GIs returned home from World War II. It's pretty cool that the "new" building is nearly 70 years old.

We begin our tour of the facility by riding the elevator to the top floors of Malt House 3 where steeping occurs, and I am happy to have Larry Truss as my guide. Larry is the malthouse supervisor, has 27 years of experience covering almost every job in the place, and has an intimate knowledge of the history, processes, and people that make the plant run. Three separate batches start each day (consisting of 62,000 pounds of barley in each of four steeping tanks); each batch totals 112 metric tons. The barley will spend just under two days in these tanks, with three alternating periods of immersion and air rest before moving to the germination box. There are 24 steeping tanks total, meaning that each day one set or the other is being emptied to get ready for a new batch. Larry pays particular attention to the steeping process. Even though temperature-controlled chilled water is used for the steep, the precise timing of the cycles needs to be monitored and adjusted to accommodate the changing environmental conditions throughout the year that influence the water uptake and respiration rates of the barley. "The first cycle is critical; over-steeped it will die, under-steeped it won't move (germinate). If you don't get it right in the steep, you can kiss it goodbye."

The fully hydrated barley moves to one of the 12 large germination beds on a lower floor where it will slowly sprout over the next 84 hours. Periodically, the turning machines make their way through the bed to prevent the rootlets from matting, to loft the bed, and to make the batch homogenous. Moist, 60°F air is constantly supplied to the sealed germination room; it will flow downward through the germinating grain bed providing oxygen and removing carbon dioxide as the barley germinates.

At the end of the germination cycle it takes about an hour for the grain to be automatically conveyed to the upper kiln. Like everything else at this malthouse, the 40-foot wide, 180-foot long kilns are industrial in size and appearance. The upper kiln will feed heated air through the bed for about 16 hours until the grain moisture has been reduced to less than 20 percent. A tipping floor allows the grain to drop to the lower deck of the kiln where it will finish drying. During the final 16-hour curing cycle, the grain bed temperature slowly increases from 140 to 180°F to finish the malt.

Rahr Malting
Shakopee, Minnesota

There are about 20 large scale malting plants located in the US and Canada. Together they have the capacity to produce three million metric tons of malt per year. The Rahr facility in Shakopee, Minnesota is the second largest malthouse in the world and is able to produce ~370,000 metric tons per year. The scale of the operations is massive; the power plant, storage, and conveying infrastructure needed to support the receiving, production, and shipping of ten plus railcars of malt every day is overwhelming.

For large commercial scale operations, the supply chain that feeds the malthouse needs to be well integrated; most enterprises of this scale have remote barley collection and storage terminals located closer to the farmland where the grains are grown. After the barley arrives at the malt house by train or by truck, it is moved to a barley storage silo where it will be held until needed. On its way to the steep tank it travels through fully-automated cleaning, grading, and weighing equipment.

Rahr has been family owned and operated since 1847, and it is evident that multiple expansions have shaped the Shakopee site since operations began at this location in the 1930s. At Rahr, there are both Saladin boxes and tower malting systems in use. Although the view of the Minnesota River Valley when standing atop the 265-foot tall, 78-foot diameter cylindrical concrete malt tower is impressive, what is happening inside on the 10 malting floors is even more so.

To start the malting process, barley is conveyed to the six open-topped cylindroconical stainless steel steep tanks located on the top floors of the tower. 2133 bu (102,400 pounds) of barley and 6600 gallons of water are added to these massive vessels. The barley soaks for 18 hours before the water is drained, and is then dropped into lower steep tanks. There is a screened drain at the bottom of each tank that retains the barley as the water runs off. Air injection nozzles at the bottom of the tank are used during the steep to periodically replenish the dissolved oxygen levels needed to combat water sensitivity in the grain. Additional water is periodically added and allowed to spill into an overflow drain, which carries away any floating debris that has risen to the surface during soaking and aeration. The steeping process lasts for a total of 40 hours before the grain drops (by gravity) to a 55-foot diameter, three-meter deep pre-germination vessel. After one day, the malt piece moves to one of the four germination floors via a shaft located in the center of the tower.

As the hydrated barley arrives in the room, it is distributed to an even bed depth of about 38 inches across a perforated metal floor. Here it will spend four days slowly growing; temperature and moisture within the bed are automatically controlled using humidity corrected, forced air ventilation that is pushed into the space above the grain. Additional "sprinkling water" is applied as needed to maintain optimal moisture levels in the piece. As the air passes through the germinating grain, it replaces the oxygen depleted by the respiring barley and removes the CO_2 that has been liberated by growth.

At the end of the germination period, the malt is removed from the bed using machinery to move it down the central shaft to a kiln floor located on the lower levels of the tower. Here, the malt is placed and leveled on the perforated kilning floor. During the initial drying phase, the incoming temperature has little effect on the temperature of air that is exhausted from the top of the bed, due to evaporative cooling.

It is only after the moisture has dropped to about 20 percent that the grain temperature begins to rise. After this "break," additional color and flavor development starts in earnest. Like most modern malthouses, the tower malting at Rahr is designed to make use of the relatively dry and warm air by moving the grain from an upper kiln to a lower kiln after the break. The hot air that is discharged from the lower kiln is reused to supply the upper deck. The cycle for emptying, cleaning, refilling, and drying on each kiln floor spans one day, so each piece of malt spends a total of two days in kilning.

The rootlets are removed as the malt exits the kiln by a deculming machine equipped with sieving cylinders. The roots are high in protein and amino acids, and are easily broken off. At most malthouses they are sold as animal feed; at Rahr they are part of the biofuel used to provide heat and electricity for the plant. The $60 million, 22-megawatt biomass fueled "Koda" project is a large scale joint venture involving both Rahr and the Shakopee Mdewakanton Sioux tribe. In addition to rootlets, the plant also utilizes wood chips, prairie grasses, and grain waste to produce electricity which is sold to the utility.

After the malt is dry it is moved to a "day bin" where it is held to allow quality control tests before being loaded into large blending/storage bins. At any one time, Rahr has 1.8 million bushels of stored barley, 3.2 million bushels of finished malt, and 536,900 bushels in process. Rahr also owns and operates a six-million-bushel grain storage facility close to the barley fields in Taft, North Dakota.

Looking over the nearly finished malt in the softly lit cavernous room, every sense is struck by the serenity of it all. Perhaps most memorable is the distinctive aroma that permeates the entire plant. These large industrial maltings might not seem as connected to the product as the smaller floor maltings operations, but the people who run them definitely have the same sense of pride in their product.

6

Malt Chemistry

"There are no applied sciences. There are only applications of science and this is a very different matter... the applications of science is easy to one who is master of the theory of it."

-Louis Pasteur

The purpose of a barley seed is to create a new barley plant. The purpose of malting is to manipulate the natural functions of the seed to create and release fermentable and nonfermentable components to make beer. The purpose of this chapter is to give a brewer a comprehensive summary of the miniature factory that changes raw materials into usable products using tiny biological machines.

The barley kernel houses the factory – it is where all of the manufacturing processes take place. The raw materials are carbohydrates, proteins, and lipids. The enzymes are the machines that transform raw materials into the essential components for making beer: starches, sugars, peptides, amino acids, and fatty acids. All of these components come together during steeping and germination to produce green malt. The final step of kilning (and sometimes roasting) is like the packaging line where the final product is prepared for delivery to and use in the brewery.

Although this book will explore the structure of the barley plant and kernel in greater detail in Chapter 8, to understand this chapter it is important to know that the food source for the kernel (the endosperm) is a tightly packed mass of starch. A thin, living layer (the aleurone layer) surrounds the endosperm. During germination, the cells in this layer generate the enzymes that break down the endosperm to release the nutrients, sugars, and Free Amino Nitrogen (FAN) that the embryo needs for growth.

Introduction to Enzymes and Modification

Structurally, an enzyme is an intricately folded three-dimensional protein molecule. Each enzyme has a specialized shape that allows it to interact with other compounds (referred to as "substrates") and cause structural changes, resulting in new "products." They are the keys that lock or unlock other molecules to facilitate biochemical reactions. Reactions motivated by enzymes can be millions of times faster than those without.

Enzymes are catalysts; they remain unchanged even as they convert substrates to reaction products. They are the engines of life that guide biochemical processes in living organisms. However, these large, complex, highly ordered, and fragile constructions can be damaged or destroyed in many ways. Excessive heat can cause the molecule to unfold (denature) thus removing its ability to facilitate the intended reaction. Acidic or basic conditions can also cause an enzyme to denature, as the internal electrochemical bonds are affected by pH.

Enzymes are conveniently named after the substrates they affect; amylases degrade amylases, beta gluconases work on beta glucans, etc. Enzymes also tend to have very specific functions and will only affect a single type of bond or substrate. For example, alpha amylase will only affect the alpha $1\rightarrow4$ bonds of glucose chains (but may do so at any location along the chain), whereas beta amylase can only cleave the $1\rightarrow4$ bond two glucose units away from the end of the chain. In the case of beta gluconases, there is a specific enzyme that can cleave the $1\rightarrow3$ bonds, and another that can cleave the $1\rightarrow4$ bond. Almost every biological compound has at least one associated enzyme, as the building and destruction of compounds forms the basis for life at the molecular level.

During malting, hydration of the kernel starts at the basal end.[*] The embryo and husk absorb water more readily than the endosperm. As the embryo hydrates, it releases hormones (including gibberellins), which awaken the scutellum[**] and aleurone layer, causing them to produce their own enzymes and begin to break down the endosperm. This process of modification starts adjacent to the scutellum at the basal end and proceeds towards the distal end. As it reacts, the aleurone layer creates beta glucanases, protein proteases, alpha amylase, and glucoamylase. The entire aleurone layer around the endosperm does not act all at once; it progresses slowly as the hormones from the embryo diffuse along the layer. As the enzymes work to break down the protein structure, the overall progression moves from basal to distal and from the outside in towards the center. Modification completes when the entire endosperm has been degraded, changing from a steely hard consistency to a mealy or mushy consistency.

Modification of the endosperm is threefold: first the dissolution and degradation of endosperm cell walls, then the degradation of the protein matrix surrounding the starch granules, and lastly the gross hydrolysis (breakdown) of starch granules.[***] The structure of the endosperm is made up of cells containing large and small starch granules, surrounded by a protein matrix. The thin, 2μm (micron) cell walls are made of beta glucans, hemicellulose, and a little bit of cellulose. Understanding how modification works requires a deeper dive into organic chemistry. The following sections will explain the molecular structures of carbohydrates and proteins. This information may be dense, but it will help explain *how* and *why* malting and mashing work the way they do.

Carbohydrates

Plants manufacture carbohydrates via photosynthesis, using water and atmospheric carbon dioxide. Plants use carbohydrates as building material and energy storage. The properties of the carbohydrate molecules depend on how the base ingredients are chemically configured. Carbohydrate molecules consist solely of carbon, hydrogen, and oxygen, and generally have a hydrogen-to-oxygen ratio of two-to-one. Both starches and saccharides (sugars) are in this group. The most basic form of these carbohydrates

[*] The basal end is where the kernel was attached to the barley plant. The distal end is the distant or outer end.
[**] The scutellum is a thin "shield" layer of cells located between the embryo and endosperm.
[***] The process is actually more complicated than that, but these three cover the basics.

is monosaccharides (single molecule sugars). Monosaccharides have the chemical formula $C_x(H_2O)_x$, where "x" is generally greater than three, but ranges anywhere from two to seven. Typically, the ratio of carbon to oxygen is one-to-one. Monosaccharides can link together (like Lego® bricks) to form larger and more complex structures called polysaccharides. These compound molecules can take on any number of forms including starches, cellulose, hemicellulose, and gums.

Yeast metabolizes digestible carbohydrates during fermentation. From a brewing perspective, carbohydrates can be broadly separated by the simple question, "is it fermentable or not?" Yeast can only metabolize six-carbon hexose sugars; they cannot ferment five-carbon pentose sugars, nor any of the three, four, or seven carbon sugars. Yeast can ferment the monosaccharides glucose, fructose, and galactose; the disaccharides sucrose and maltose, and in the case of lager yeast, the trisaccharide maltotriose. Anything larger (like maltrotetraose) is considered a dextrin (or its formal name, oligosaccharide), and is unfermentable. Although there is overlap, yeast preferentially uptake glucose over fructose. To metabolize sucrose, yeast first breaks it down into glucose and fructose outside of the cell using extracellular enzymes. In the presence of glucose, the transport mechanism for bringing maltose and maltotriose into the cell is repressed, and the yeast consumes the easily available monosaccharides before spending energy on larger molecules.

Sugars

The most important sugars to brewers are six-carbon hexoses with the chemical formula $C_6H_{12}O_6$ (See Figure 6.1). These primary building blocks of carbohydrates are very important in malting and brewing.

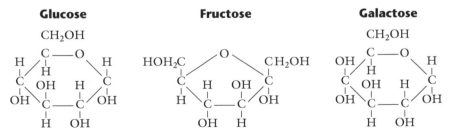

Fig. 6.1: Structure of Glucose, Fructose, and Galactose. ©John Palmer

Glucose is the basic food source for all life on earth; it can be absorbed and directly metabolized by living cells, including yeast. Other monosaccharide building blocks include fructose and galactose, both of which are structural isomers of glucose. Fructose (laevulose or fruit sugar) is, as the name suggests, commonly found in fruits and can be derived from corn as well. Galactose is a component of lactose (milk sugar). Lactose[*] is a disaccharide composed of one glucose and one galactose. Sucrose (table sugar) is composed of glucose and fructose.

Sugar Structure and Atomic Bonding

The structure of a sugar like glucose can vary, existing in either a linear or cyclical form because of its atomic bond arrangement. As you may have retained from high school chemistry class, when two atoms share electrons, they bond together. If multiple electrons are shared between a pair of atoms they are "double bonded." Atoms have a specific number of bonds that are made in a normal state. For carbon, this number is four, for oxygen it is two, and for hydrogen it is one. Water is made by utilizing the two bonding sites of oxygen with the one from each of two hydrogen atoms. Two double bonds exist in CO_2.

H_2O CO_2

Fig. 6.2: Water has two single bonds. CO_2 has two double bonds.

Sugar crystals are highly ordered arrangements of individual sugar molecules in their cyclical form. In solution, some molecules will transition to their linear form. The change of sugar from linear (acrylic) to cyclical form rearranges bonds but does not change the number or essential arrangement of atoms in the molecule.

---

[*] Lactose is not fermentable by yeast, but galactose is, if lactose is hydrolyzed by enzyme additions. Galactose is typically not found in wort, even though it is present in minute concentrations during germination.

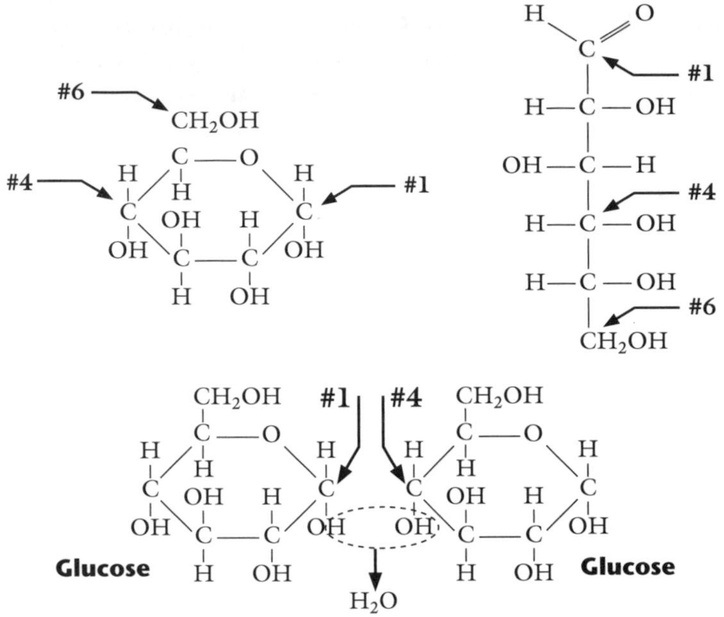

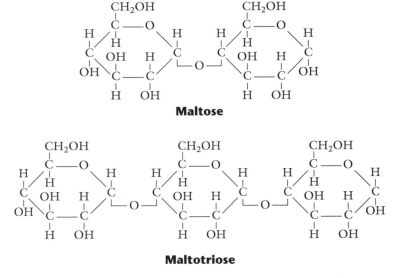

Fig. 6.3: Glucose in cyclical and linear forms. Carbon atoms in sugars are numbered starting at the end with the double bonded oxygen. ©John Palmer

Maltose

Maltotriose

Fig. 6.4: Diagram showing the bonding for maltose and maltotriose. The top diagram shows how the carbon 1→4 bond occurs via the loss of a water molecule, leaving the oxygen bonded to both carbon atoms. ©John Palmer

Molecular bonding of monosaccharides requires the loss of one oxygen atom and two hydrogen atoms (more popularly known as a water molecule). Organic chemists identify points of bonding by numbering the carbon atoms in the linear molecular form. "Carbon #1" is located at the end closest to the double-bonded oxygen (see Figure 6.2). Attaching carbon #1 of a glucose molecule to carbon #4 of a second glucose molecule forms maltose. Attaching another glucose using a different 1→4 linkage yields maltotriose (see Figure 6.4). Carbon 1→4 bonding of glucose and fructose creates sucrose. Glucose and galactose bonded in 1→4 configuration yields lactose. Glucose is also able to form 1→6 linkages, and if a 1→6 bond is present in an otherwise 1→4 chain, the chain has the ability to branch. The bonding location is important for understanding starch structure, which will be covered in the next section.

Raffinose, another tri-saccharide, is a composed of galactose, fructose, and glucose. Raffinose accounts for about 25 percent of the sugars present in a kernel. More than 80 percent is present in the embryo, and is metabolized rapidly in germination. The human body does not digest this sugar well; when it is finally fermented by bacteria in the lower intestine, it causes flatulence. Foods high in raffinose include beans, cabbage, brussels sprouts, and broccoli.

A Note on Sweetness

When most people think sugar, they think sweet. Different types of sugars have different perceived levels of sweetness. Common table sugar, sucrose, is the reference standard. Glucose (also known as dextrose and blood sugar) is only 80 percent as sweet. Fructose is 1.7 times as sweet as sucrose. Maltose and lactose have relative sweetness values of 45 percent and 16 percent when compared to sucrose. Non-carbohydrate sweeteners, like Splenda, can have values up to 600 times that of sucrose. Many of these compounds were accidentally discovered when laboratory workers absentmindedly put their fingers in their mouths. Other naturally occurring compounds (such as glycyrrhizin which is found in licorice root and sometimes used as brewer's licorice) are also perceived as sweet.

Invert Sugar

One fascinating property of sugar solutions is their ability to rotate polarized light. A sucrose solution is dextrorotatory, which means it causes light passing through it to rotate to the right (clockwise). If sucrose is split into its constituent parts, fructose and glucose, the resultant solution is laevorotatory (left turning/ counter-clockwise). This is because fructose is more powerfully laevorotatory than glucose is dextrorotatory. This inversion of optical polarity is a good indicator of the breakdown of sucrose during the production of "invert sugar," a traditional ingredient used in some British beer styles.

Starches

Starches are very long polymers of glucose. If only 1→4 linkages are present, the molecule will form an unbranched chain (like a string of beads) called amylose. In barley malt, these chains are roughly 2000 glucose units in length. Amylopectins are composed of amylose chains but also contain 1→6 linkages roughly every 30 sugar units. They are highly branched molecules that can be larger than 100,000 glucose units in size. Structurally, amylose forms in a helix; long, tight spiral shapes which are both denser and harder to breakdown than amylopectin.

The endosperm can be up to 80 percent of dry grain weight. The starches themselves (contained inside the endosperm) can be up to 65 percent of the dry weight. As noted earlier, the internal structure of the endosperm is a mixture of large and small starch granules embedded within a protein matrix. Roughly 25 percent of the starch mass exists as amylose; the remainder is amylopectin. Although the small granules comprise 80 to 90 percent of the total starch *count*, the large starch granules contain about 90 percent of the *weight* of starch in the grain. The small granules measure roughly 5 μm in diameter while the large are about 25 μm.

While these numbers may not seem terribly riveting, they are quite literally what make malting and mashing of barley possible. During malting, the numerous small granules (with their high surface area) are readily and almost completely degraded by the alpha amylase and glucoamylase enzymes, producing glucose for the growth of the acrospire. The large starch granules on the other hand, which contain the majority of the total starch, are only poked at by the enzymes, leaving the majority available for the brewer.

Another brewer-applicable property of amylose and amylopectin is the ability to bind with iodine to produce a blue color. The spiral form of amylose traps iodine inside its coil and makes the blue iodine clearly visible. Amylopectins don't have this trapping ability and iodine reacts weakly with it (which is visible as a reddish color). The large granules generally have a higher percentage of amylose than the small granules, and the total percentage of amylose in barley starch is estimated at about 30 percent. The starch test that brewers use to gauge whether saccharification is complete utilizes the roughly 25 percent of the amylose contained in the large starch granules.

Isomers – Not Just an Evil Twin

The three dimensional structure of organic compounds is key to their chemical and physical properties. Two organic compounds can have the same atomic formula (such as $C_6H_{12}O_6$) but have different properties; these chemical clones are called isomers. There are two main types of isomers: structural and stereo. Structural isomers have atoms bonded in a different order, such as ABCD vs. BCAD. Stereoisomers have the same general bonding order but are arranged differently, such as $_ABC_D$ vs. $_ABC^D$.

The Other Substances - Non-starch Polysaccharides

The endosperm houses other non-starch polysaccharides, which are primary components of the cell walls that enclose the protein matrix of starch granules. Pentoses join the hexoses to build these structures. As mentioned earlier, the cell walls of the endosperm are primarily composed of beta glucans, hemicellulose, and cellulose. Beta glucans make up the majority; about 75 percent. The beta glucans are made from beta-glucose molecules (as opposed to alpha-glucose) and are attached to each other by both β 1→3 and β 1→4 bonds (see Figures 6.5 and 6.6).

Degradation and dissolution of beta glucans by the glucanase enzymes into oligosaccharides (and eventually glucose) opens the endosperm to further protein degradation and starch conversion. Two primary endo-β-glucanase enzymes take part in this task. Both act on the β 1→4 bonds that are adjacent to the β 1→3 bond on the non-reducing end of the chain, producing three or four glucose unit oligosaccharides (glucans, not dextrins) that are eventually broken down into glucose by other enzymes. The endo-β-glucanases are not very heat stable, and can be denatured in fewer than five minutes at 150°F (65°C). An "endo" enzyme works within the chain, while an "exo" works on the end of the chain.

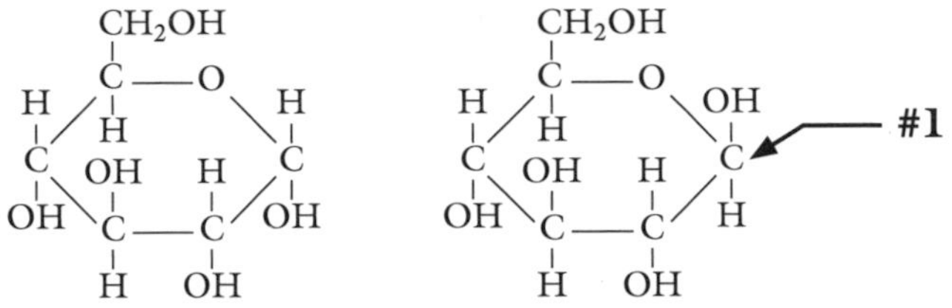

α Glucose   β Glucose

Fig. 6.5: A comparison of the Haworth diagrams for alpha and beta glucose. The molecules differ by the rotation of the hydroxide and hydrogen around carbon #1.
©John Palmer

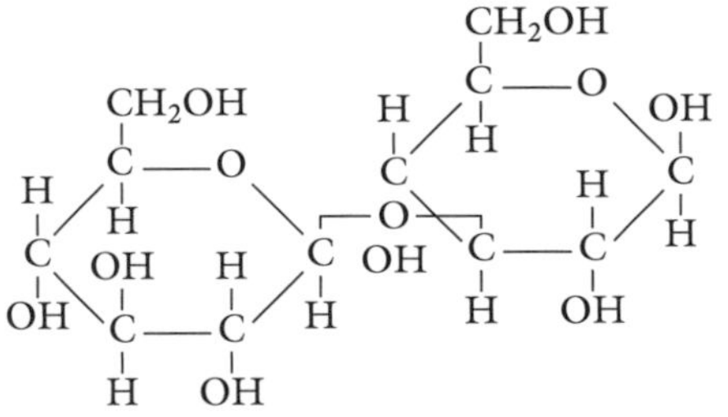

β Glucan 1→3

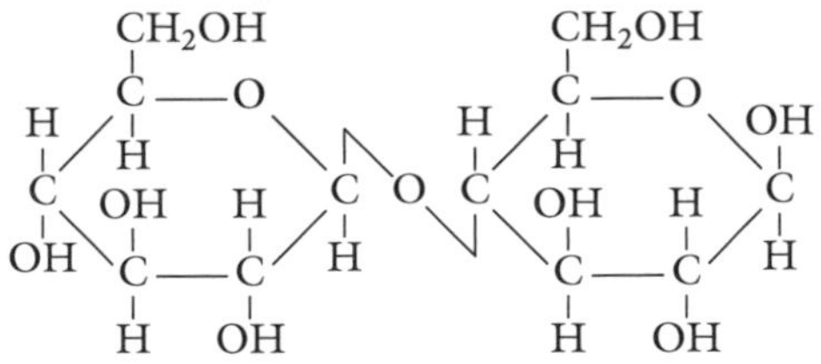

β Glucan 1→4

Fig. 6.6: Haworth diagrams of the beta 1→3 and beta 1→4 bonds that form the glucose chains for beta glucan molecules. The bonds are glycosidic (the same as in starch chains), but the difference in carbon attachment points, and the alpha versus beta glucose structure, prevents amylase enzymes from acting on them.
©John Palmer

Hemicelluloses are highly branched molecules composed of many different types of sugars. Barley cell walls are about 20 percent hemicellulose; mainly arabinoxylan. Hemicelluloses (also known as pentosans) are long chain polymers composed mostly of monosaccharide pentoses like arabinose and xylose.

Ferulic Acid

Ferulic acid is an important component of barley and wheat. It is concentrated in the cell walls of the seed coat and aleurone layer where it is cross-linked with arabinoxylan (a hemicellulose) to help hold the cell walls together. Ferulic acid (3-methoxy-4-hydroxycinnamic acid) is the precursor used by yeast to produce 4-vinyl-guaiacol (4VG), which is the clove-like aroma found in wheat beers.

Cellulose comprises only about two percent of the endosperm cell wall, but it provides structural rigidity. Cellulose is a straight chain polymer of $\beta1\rightarrow4$ bonded glucose, but unlike starch or hemicellulose, it does not coil or branch, and the molecule has a less amorphous, more crystalline structure. The multiple hydroxyl groups (^-OH) on the glucoses of one chain form hydrogen bonds with oxygen atoms on the neighboring chain, holding them firmly together side-by-side, forming microfibers with high tensile strength. These microfibers bolster the cell walls by forming a structural composite with the beta glucan and hemicellulose matrix. Phenolic acids (such as ferulic acid) act as a glue or resin that hold this composite together by being able to molecularly bond and bridge the components. Cellulose itself does not degrade in the presence of enzymes during the malting or brewing process but instead passes through and is still intact in the spent grain.

Phenols are hydrocarbons with a ring shaped molecular structure. Unlike saccharides, phenols lack oxygen atoms in their structure. The most basic type of phenol is a hydrocarbon ring (C_6H_6) plus one oxygen, which forms a hydroxyl group (^-OH), ultimately creating C_6H_5OH. Many phenols have an aroma and are known as aromatic hydrocarbons. When multiple phenols join together to make larger structures, polyphenols are created.

Polyphenols are present in the husk and cell walls of barley and are not directly affected by enzymes. Tannins are a subset of polyphenols containing a large amount of hydroxyl ($^-$OH), carboxyl ($^-$COOH), and other active groups that allow them to readily bind with proteins. They are important to brewing for haze formation, but also contribute astringency to beer. For example, chill hazes are caused by the temporary linkage of polyphenols to proteins in cold conditions. When the beer is warmed the linkage is broken and the haze disappears. If oxygen is present, over time, these bonds can polymerize and form larger molecules, causing the haze to become permanent. This is an example of how raw material choices, brewing technique, and oxygen control all contribute to the quality of a finished beer. Polyphenols also provide some antioxidant properties.

Proteins

At the most basic level, the difference between a carbohydrate molecule and a protein molecule is the addition of a nitrogen atom (or several of them). Nitrogen can make three bonds as opposed to carbon's four. Ammonia is a common molecule with a sharp aroma that is sometimes used in breweries as an environmentally benign refrigerant. Ammonia has the formula of NH_3, and if one of the hydrogen atoms is removed, an amine group ($-NH_2$) remains. When added to hydrated carbon compounds, amino acids, and ultimately proteins, can be created.

Amino acids are so named because they have both an amine ($^-$NH2) and a carboxylic acid ($^-$COOH) in the molecule. These "functional groups" are located at one end of the molecule in the 23 basic amino acids that build larger peptides and proteins, and typically have the generic formula $H_2NCHRCOOH$ (where R is an organic substituent). The side chain "R" can vary in size, structure, and composition. Amino acids have the ability to bond together at the functional group end. A "peptide bond" results when a hydrogen atom is lost from the amine group of one amino acid and the ($^-$OH) is removed from the carboxyl group of another.

Molecular Building Blocks

Biochemical molecular structure uses simple building blocks to make up larger structures. Proteins are structured in the same way as monosaccharide sugars that make up larger polysaccharides, which in turn make up large starch molecules. Amino acids make up (in ascending size) peptides, polypeptides and proteins. Technically, peptides are small proteins or protein segments that are larger than amino acids, but smaller than the polypeptides and proteins that the protease-class enzymes affect.

Living organisms use 20 different amino acids to construct larger structures. A polypeptide results when several amino acids join together using peptide bonds. Proteins are constructed of one or more polypeptides and have biochemical functionality that comes from their very specific physical structure. Proteins can be differentiated based on their solubility in the laboratory: albumins are soluble in water, globulins are soluble in dilute salt solutions, prolamins are soluble in alcohol solutions, and glutelins are not soluble in any of the above solutions. Cereal chemists organize barley proteins into two main groups: storage and non-storage, based on their location and function within the kernel. Storage proteins serve as peptide and free amino acid (FAN) reservoirs for the embryo, and include hordeins (prolamins) and globulins. Non-storage proteins are the structural proteins and enzymes, which include albumins, glutelins, and globulins. Globulin proteins exist in both groups because solubility is not very indicative of function.

During malting, the endosperm's protein matrix is hydrolyzed into polypeptides, oligopeptides, and free amino acids. These proteins are a mixture of hordeins (prolamin proteins) and to a lesser extent, glutelin proteins. The hordeins are the primary component of the protein matrix surrounding the starch granules, and it is the breakdown of this matrix during germination that provides the vast majority of FAN to the wort. In a study by Lekkas,[1] 28 different two-row malt samples of six varieties all demonstrated that at least 70 percent (in many cases up to 90 percent) of the FAN in the wort had been created during malting.

The matrix also contains glutelin, which seems to be the source of the larger soluble peptides that survive into the wort, and may yield FAN during mashing. The non-storage proteins are the source of the enzymes that are present in barley before malting, such as beta amylase, and albumins such as protein Z (which is a primary foam former in beer).

There are two main categories of enzymes that degrade endosperm protein during germination. The first are endoproteases and endopeptidases, which act to break up the protein molecules from the inside. There are at least 40 such enzymes involved in this stage.[*] The second group is exoenzymes (such as carboxypeptidase) that produce individual amino acids from the carboxyl end of the peptide chain.

High Protein vs Low Protein Malts

Most brewers do not want to brew with malt made from barley with high protein content for several reasons:

- Higher protein levels tend to produce more beer haze

- Higher protein levels provide more nutrients for spoilage microorganisms

- Mixed lots of higher and lower protein malt can cause inconsistent brewing results

- Higher protein levels means less fermentable extract per pound of grain, which means higher grain costs

In response to drought, barley develops higher protein levels. Realistic brewers know that they have to use the barley that nature gives them or not brew at all, which for most breweries is not a very viable option.

Glutens in Barley

There are several conditions that can cause a negative physical reaction when someone drinks beer. An allergic reaction to barley, may or may not be associated with gluten, as the grain contains two dozen different allergens. People can be allergic to barley just as they can be allergic to wheat, cats, eggs, and peanuts. However, people can also be specifically sensitive to gluten, which can manifest in several ways. One is a straight allergic reaction: watery eyes, runny nose, and respiratory problems. A second is "gluten rash;" a type of dermatitis that is caused by an autoimmune response. The third condition, Celiac disease, is a very serious autoimmune disorder that damages the small

---

[*] Bamforth, *Scientific Principles of Malting and Brewing*, p. 54.

intestine, prevents it from absorbing nutrients, and can cause melanoma and other cancers. Lastly, gluten "sensitivity" or "intolerance" will cause symptoms very similar to Celiac disease but without the damage to the small intestine. Sufferers of Celiac disease have an immunoreactive response to gliadin, which is a prolamin in wheat that combines with other wheat proteins to form gluten (which is what gives bread dough both elasticity and structure).

Gliadin is closely related to hordein (present in barley) and secalin (found in rye). All are considered prolamins (storage proteins that contain the amino acid proline) and are present in the grains of grasses. This is the reason that even though barley doesn't technically contain gliadin, beer might still be a problem for people with gluten sensitivity. A proline-specific endo-protease enzyme that is reported to completely breakdown the specific protein sequences in barley that react with standard gliadin test methods can now be added to wort and beer. At the time of this writing, the USDA has not approved labeling beverages using this enzyme as "gluten-free," because it is not definitely known that these prolamins are the only factor for the disease, and there is still debate about what level of residual prolamin (as measured by the gliadin test) is tolerable.

Lipids

Waxes, fats, fatty acids, vitamins, and sterols (such as cholesterol) are all in a class of compounds called lipids. All lipid molecules have a hydrophobic portion that does not associate well with water, and another portion that does. For example, fatty acids are long chain hydrocarbons with a carboxylic acid (^-COOH) end. The acid end is polar, so it is hydrophilic (water loving). The end result is a molecule that causes one end to attract water and another that repels it, like a magnet. Lipids can therefore act as a bridge between polar and non-polar molecules, and participate in many biochemical reactions in the cell. Fatty acids are used biologically both in the formation of cell walls and for energy storage. The fatty acid content in barley is roughly 58 percent linoleic acid, 20 percent palmitic acid, 13 percent oleic acid, eight percent linolenic acid, and one percent stearic acid.

Lipids in barley can be divided into starch and non-starch lipids, much like protein classification. About two-thirds of the total lipids are stored in the endosperm and aleurone layer, and the other fraction is found within the embryo. Most (~75 percent) of the lipids in barley (such as glycerides) are non-polar. The polar lipids include the glycolipids and phospholipids (which in turn include fatty acids). Lipids are most notorious for contributing compounds that cause beer to go stale. Enzymes such as lipoxygenase oxidize fatty acids into hyperperoxides, and add carbonyl compounds to beer (e.g. aldehydes). Brewers recirculate wort (vorlauf) until it is clear to remove this lipid material to prevent premature staling. While some lipids are needed for yeast nutrition, the quantity is quite small, and brilliant wort contains an adequate amount.

Browning Reactions in the Kiln and Kernel

Kilning or roasting is the last phase of malting, and when the final changes happen inside the kernel. Most of the aromas associated with malt are the result of thermal reactions. Technically, caramelization is the thermal degradation of sugars which leads to the formation of both volatiles (caramel aroma) and brown-colored products (caramel colors). It is similar to Maillard reactions in that it is a non-enzymatic browning reaction, but caramelization occurs by pyrolysis (thermochemical degradation in the absence of oxygen) instead of by chemical reaction. Caramelization is catalyzed by an acid or a base, and generally requires temperatures above 120°C and a pH between three and nine. The high temperatures preclude the presence of water unless the system is under high pressure. In normal malting and brewing practice these conditions are rarely present, so how do these malt flavors arise?

The History of Maillard

The discovery of an alternate set of reaction pathways involved both a brewing scientist and a tenacious French chemist in the early 20th century.

Arthur Robert Ling was a highly regarded brewing chemist known for his work on starch, sugars, and brewing. As editor of the Journal of the Institute of Brewing, Lecturer on Brewing and Malting at the Sir John Case Institute in London, and even Vice-President of the Society of Chemical Industry, he was a brewing industry rockstar. In 1908, he presented novel findings at a meeting in London where he described the formation of color compounds. "When these amino-compounds produced from proteins are heated at 120–140°C with sugars such as ordinary glucose or maltose, which are produced at this stage of process, combination occurs."

Intrigued by Arthur's findings, thirty-year-old French chemist Louis Camille Maillard did substantial work on the reaction products created when amino acids and sugars were heated together. He first published his findings in 1912 and these reactions bear his name today. They describe and explain the characteristic flavors of chocolate, roasted coffee, bread crusts, maple syrup, soy sauce, cooked meat, and malt.

Maillard reactions can produce many of the same flavors and aromas as caramelization reactions, but at much lower temperatures. The mechanism for the reaction starts with the carbonyl groups in simple sugars and the free amino groups in amino acids. In addition to the common bread crust, caramel, cocoa, and coffee flavors, less pleasant aromas and flavors such as burnt, onion, solvent, rancid, sweaty, and cabbage can be created.

The reaction occurs in three steps. In the first step, an amino acid and a sugar combine (with the loss of a water molecule) to form an unstable compound. During the second step, this unstable compound undergoes Amadori rearrangement (an isomerization reaction) to form a ketosamine (combination of a ketone and an amine). During the final step, the ketosamine undergoes further transformation (via one of three different pathways) to produce one of three different products.

The first pathway further dehydrates the ketosamine, resulting in compounds like those formed in true caramelization reactions. The loss of three water molecules and additional reaction with amino acids characterize the second pathway, which results in the creation of large, colored polymeric compounds called melanoidins. In the third pathway, an intermediate product such as diacetyl is formed, which then undergoes Strecker degradation (conversion of an amino acid to an aldehyde) to form highly flavor active heterocyclic compounds (including pyrones like maltol and isomaltol), as well as furans and furfurals.* Stronger flavored nitrogen heterocyclic compounds, including nitrosamines, tend to be more abundant in malt that has been kilned or roasted above 350°F (180°C).

---

* Furfurals are used as a measure of heat stress on beer. Beer that has been scorched in the kettle will become stale faster than normally processed beer. The Indicator Time Test is one laboratory measure of the heat stress that beer has undergone. ITT measures furfural levels. Aged oxidized beer also has higher levels of these compounds.

The conditions necessary for Maillard reactions can be found both in the malt kiln and brewhouse. In fact, the majority of color in pale North American lager beer is generated in the brew kettle via these reactions. Roughly 10,000 distinct compounds (like maltol, the characteristic aroma of Munich malt) originate during brewing. The complexity of these formation pathways is such that the individual products cannot be precisely controlled, but maltsters can steer the flavors in a general direction. Although caramel flavors can be created in the kettle, they occur mainly due to Maillard reactions.

Diastatic Power in Malts

The only amylase that exists in barley prior to malting is beta (β) amylase, which is present in most, if not all, of the tissues of the barley plant.[2] Beta amylase exists in both free and bound form, meaning that some of it is bound to other compounds (such as protein Z) and is only released later during germination or mashing, through proteolytic action. Alpha (α) amylase is produced in the aleurone layer during germination, along with glucoamylase (also known as alpha-glucosidase) and limited amounts of dextrinase. The question is: how much enzyme is left after kilning and available to the brewer?

The diastatic power in a maltster's Certificate of Analysis (COA) is a measure of a malt sample's ability to produce sugar from a known quantity of a standardized starch solution. This test can be conducted several ways, but the baseline is a wet chemical method that can take all day to complete. An automated flow analysis (ASBC MOA Malt-6C) is more commonly used for production testing. These tests do not measure the actual amount of the enzymes in the malt sample, but instead measure the production of sugars generated by the enzymes in the malt sample. The original test was proposed by Carl Lintner in 1886 and has since been modified in many small but significant ways, including improvements to the solutions used for the extraction of the enzymes from the malt. The official unit for diastatic power is no longer properly referred to as °Lintner, but as simply "diastatic power, degrees ASBC" (although °L is still used in common parlance). In Lintner's original method, a malt had a diastatic power of 100°L if 0.1mL of a five percent infusion, acting on a starch substrate under fixed conditions, produced sufficient sugars to completely reduce five mL of Fehling's solution. Fehling's solution is mixture of copper sulfate, potassium sodium tartrate, and sodium hydroxide that changes color to identify the presence of monosaccharides in the solution.

A second standard test on the COA uses a special starch substrate that has already been completely converted by laboratory beta amylase to measure the dextrinizing power of alpha amylase. An α amylase unit (or dextrinizing unit, DU) is defined as the quantity of α amylase that will dextrinize soluble starch in the presence of excess β amylase at a rate of one gram per hour. This test also measures activity from debranching enzymes.

Enzyme Action

During malting, the endosperm is fully modified and the starch granules have been exposed. In effect they are pin-holed by alpha amylase. In the mash, the milling of the grist greatly increases the surface area of the endosperm on which enzymes act, and all four amylase enzymes take part in producing the sugars that make up a typical wort of 41 percent maltose, 14 percent maltotriose, six percent maltrotetraose, six percent sucrose, nine percent glucose and fructose, 22 percent dextrins, and two percent non-starchy polysaccharides (hemicelluloses).[3]

α and β amylase are the two most well-known enzymes in brewing. Although both will attack starch at a 1→4 linkage (and break it down into smaller sugar units) each has a different mode of action, which dramatically affects the types and proportions of sugars produced in the wort. Alpha amylase is an endo-amylase and can hydrolyze the α1→4 bond anywhere except within one glucose unit of an α1→6 bond. Beta amylase is an exo-amylase and can only act of the end of a chain, within 3 glucose units of an α1→6 bond.

The amylase enzymes can be thought of as choppers or nibblers. Alpha amylase breaks linkages at random locations and this "chopping" action results in the production of a wide range of sugar chain lengths. β amylase acts by "nibbling" off maltose from the non-reducing end[*] of larger carbohydrate units. Yeast can only utilize glucose, maltose, and (in the case of lager strains) maltriose from the malt starch, so any larger sugars will pass through to the finished beer. These larger sugars will ultimately be perceived as sweetness, flavor, and body.

The various amylase enzymes work together in the mash to optimize saccharification. In general, higher β amylase activity will create more

---

[*] The reducing end of a sugar is the end with the aldehyde functional group (⁻CHO).

fermentable sugars, which will result in a drier beer. Acting alone, α amylase would create wort containing less than 20 percent fermentable sugar. Adding β amylase raises the fermentability to 70 percent. Adding the de-branching enzyme to the mixture will raise fermentability to 80 percent. Optimal conditions for β amylase are roughly 131°F (55°C) and 5.7 pH. In contrast, α amylase works best at 149°F (65°C) and 5.3 pH. Synergistically, they work best together in between those conditions. Glucoamylase plays only a small role during mashing, given the relatively low concentration of glucose in wort.

Mashing conditions can favor the activity of one amylase over the other, so control of mash parameters is essential to the production of consistent wort. Enzyme activity is governed by reaction kinetics. Most chemical reaction rates double with each 18°F (10°C) increase in temperature. Enzymes are destroyed above specific temperatures, and there are limits that cannot be exceeded if enzyme activity is to be maintained. For example, β amylase will be destroyed through denaturation above 155°F (68°C). Physical parameters in the mash such as pH, as well as thickness or dilution, also influence relative activity. Ions (such as calcium) also impact enzymatic activity.[*] The varying levels of enzymes within different malt lots complicate an already complicated biochemical system.

Although base malts can supply enough diastatic power to convert significant quantities of adjunct brewing materials into fermentable sugars, the production of very dry beers may require the use of external enzymes. Commercial preparations can be derived from a variety of sources, including fungus and bacteria. These products can be used in the mash or during fermentation to increase the percentage of fermentability. Scientists continue to develop enzyme products that allow the use of unmalted barley for the production of beer. Enzyme treatments that mimic natural carbohydrate and protein breakdown could significantly reduce the time and energy inputs to the brewing process. However, it is reasonable to expect that beers made from such a process would present a different flavor profile from a traditionally malted and mashed beer.

Conclusion

Anna MacLeod, the well-respected Professor of Brewing and Distilling at Heriot-Watt University in Edinburgh, Scotland once described malting as

---

[*] For example, alpha amylase cannot function in the absence of calcium. Fortunately, malt contains a fair amount of it, typically in the neighborhood of 35 mg/l in a 1.040 (10°P) wort.

"a process which allows 1) the optimal development of hydrolytic enzymes by the aleurone cells of barley and 2) the controlled action of some of these enzymes to eliminate structural impediments to the subsequent easy and complete extraction during mashing." Malting and brewing, at a most basic level, involve the chemical and biochemical manipulation of carbohydrates via a complex and varied set of factors. Although it has been demonstrated for thousands of years that comprehensive knowledge of the molecular details is not needed to brew good beer, the substantial work of researchers makes it possible to bring an unprecedented level of control for the modern maltster and brewer in search of brewhouse excellence.

References

1. R. Leach, et al, "Effects of Barley Protein Content on Barley Endosperm Texture, Processing Condition Requirements, and Malt and Beer Quality", *MBAA Technical Quarterly*, 39(4) (St. Paul, MN: MBAA, 2002) 191-202.

2. J.S. Hough, et al., *Malting and Brewing Science*. 2 vols. (New York: Chapman and Hall, 1982).

7

Malt Family Descriptions

The task of classifying hundreds of varieties of malts into neatly ordered style categories is as daunting as trying to classify the thousands of beers made from them. It is possible to broadly separate malts based on process (kilned, caramelized, roasted), enzymatic activity (can they self-convert?), or even the color of the wort that they produce, but even similar products from different producers can have overlapping qualities and unexpected differences. When coupled with batch variation and fluctuations in raw materials, categorization can be a herculean labor.

When writing or revising the grist for a recipe, the brewer should always taste the malt. There is no substitute for the direct sensory experience gained from chewing malt. Tasting a blend of the individual grist ingredients mixed together in rough proportion gives a far closer approximation of what the finished beer will taste like than any amount of written description. Tweaking the recipe before brewing begins requires little effort or commitment, and can be vital to the success of the beer.

The listings that follow are roughly grouped by type (traditionally processed malts, caramel malts, drum roasted malts, malt made from other grains, and malts made with special processes) and arranged from light to dark within these categories.

Fig. 7.1: Malt wheel showcasing the wide range of colors achieved during the roasting process of malting. Reproduced with permission. ©Thomas Fawcett & Sons, Ltd.

Author's note: As this chapter came together, I took the opportunity to have Matt Brynildson, Brewmaster of Firestone Walker Brewing Company, share some of his thoughts on malt varieties. Matt, in addition to being a fantastic brewer, is also a close friend, and we have toured many, many breweries around the world together. I love to avail myself of the opportunity to discuss brewing ideas and best practices with him.

Standard Processed Malts

Malts in this group are produced using standard steeping, germinating, and kilning techniques. All are somewhat pale and contain sufficient enzyme potential to convert their own starches. Standard, very pale malt is sometimes referred to as "white malt" and is used as the base material for producing roasted malts like black or chocolate.

Pilsner Malt

Color Range: 1.2–2 SRM

Pilsner malt is a base malt designed for very pale, all-malt beers. Traditional Pilsner malt production includes the use of low protein two row malts, lower modification during germination, and low temperature, high airflow kilning. This malt should be very pale in color with moderate enzymatic potential. According to Brynildson, Pilsner malt has a distinctive flavor: a little green, with the smell and taste of fresh wort, which are particularly evident in European Pilsner style beers, like Bitburger or Warsteiner.

Dimethyl sulfide (DMS)—which has a cooked corn or cabbage flavor—is present in all malt. The precursors S-methylmethionine (SMM) and dimethyl sulfoxide (DMSO) are created during malting, but removed at higher kiln temperatures. Because Pilsner malt is kilned at a low temperature it retains this flavor potential, which is considered acceptable at low levels in some beer styles, like German Pilsner.

Pale Malt

Color Range: 1.6–2.8 SRM

Pale malt is a generic term covering a broad class of light colored base malts. For North American maltsters, this malt is produced with adjunct brewing needs in mind. High enzymatic potential and FAN makes this malt particularly well-suited for rapid carbohydrate conversion and proper yeast nutrition. The very high enzyme potential of pale malts can make fermentability control difficult in all-malt beers as mash conversion can be almost instantaneous. For maltsters in other areas of the world, "pale" malt has moderate modification and enzyme potential. The term "lager malt" may also be used to describe these malts. In comparison to Pilsner, pale exhibits a deeper malt flavor.

Pale Ale Malt

Color Range: 2.7–3.8 SRM

Pale Ale malt is a base malt produced specifically for use in English-style pale ales. These are well-to-highly modified, tend to be darker than standard pale malts, and are optimized for use in single temperature infusion mashes. They have an evident, but not excessively pronounced malty flavor, with notes of biscuit or toast. The higher kilning temperatures used to produce these malts result in low DMS/DMSO potential. Maritime growing conditions of the UK produce barley that is particularly well suited to making pale ale malts. The plump grains respond well to the full modification schedule in germination, and result in malt that responds well to single temperature infusion mashing. These malts provide a flavor backbone that gives structure and allows for the successful production of low gravity but full flavored ales that typify English Cask Ale, like Timothy Taylor's Landlord Bitter.

Vienna Malt

Color Range: 2.5–4.0 SRM

Vienna malt imparts a rich orange color to beer. The flavor of traditional Märzen beer comes from liberal use of this malt. Vienna malt has sufficient enzymatic power to convert up to 100 percent of the grist. In contrast to most crystal malts, use of Vienna results in a beer with a refreshing, dry finish. The flavor is slightly toasty, slightly nutty, and pairs very well with spicy, noble hops. Although Vienna malt has significant flavor complexity, it is not cloying or overbearing when mashed sufficiently. Vienna malts make beer that can be drunk readily in one liter glasses and leaves the drinker ready for more, thus explaining the appeal (and longevity) of the Munich festivals.

Munich Malt

Color Range: 3–20 SRM

The Munich malt category covers a broad range of colors, with lighter versions having a restrained but often elegant character. Enzymatic potential is low but still sufficient to convert a mash when used at 100 percent. Many brewers will add a small quantity of Munich to the grist to fill out the malty profile of their beers. Munich is typically the flavor that comes to mind when drinkers think of maltiness. Brynildson is a huge fan of small Munich malt additions in virtually all pale ale recipes. Color choice of Munich can have important consequences for the final beer. The lighter versions tend to be more refined and subtle, while the darker versions have more heft and authority.

Melanoidin Malt

Color Range: 17–25 SRM

Melanoidin malt has a sweet, honey-like flavor. Although some variants have sufficient enzymes to convert up to 100 percent of the starch, it is more commonly used at lower concentrations as the flavor can become too pronounced when it is the primary malt. Also known as Honey malt or Brumalt, it is made by reducing airflow during the final stages of germination. It is kilned at lower temperatures to promote self-conversion that leads to elevated levels of Maillard reaction products. Melanoidin malts have low astringency and seem to have a greater flavor than the simple color potential would suggest. Many brewers consider this malt "Super-Munich" as it shares some flavor notes with Munich malt. Brynildson finds that an addition of up to 10 percent produces a pronounced honey aroma in beer.

Caramel Malts

Malts in this group are produced by loading green malt (steeped and germinated but not kilned) into a drum roaster. They are "mashed in the husk" and subsequently roasted.

Special Glassy Malts

Color Range: 1–12 SRM

Special Glassy malts are made using low temperature and high moisture to produce pale malt with a glassy endosperm. These malts do not have enzymatic potential and are used instead to enhance head retention, add body, or impart sweetness to beer. They are sold under a variety of proprietary trade names such as Cara-pils or Carapils. Also known as "dextrin malts," they can be used as body boosters. In the US, "Carapils" is a trademarked name owned by Briess Malt & Ingredients Company, but in the rest of the world that name is used by Weyermann Specialty Malting Company.

Caramel/Crystal Malts

Color Range: 10–200 SRM

Caramel (also known as Crystal) malts are made by raising the temperature of green malt to turn starches and proteins into the sugars and amino acids needed for Maillard reactions. Determining the best caramel malt for a given beer often involves balancing color/flavor impacts. To reach a specified color depth, a greater percentage of lighter colored malt is needed when compared to darker malt. This type of malt can be made in a standard kiln or with a drum roaster, and there can be significant flavor difference depending on the method of manufacture.[*]

Crystal malts contribute to the flavor and color of the beer equally. Sensory evaluation of the malt is the key to understanding how it will express itself in a beer. Light-to-mid color crystals are often used when crystal malt flavor is the desired main flavor for the beer. These 20 to 60 Lovibond malts tend to have a cleaner flavor that allows them to stand out as main actors on the aromatic stage of beer flavor. Darker crystal malts are useful but are generally used for the supporting flavors they contribute to beers.

Lighter caramel malts seem to be more forgiving and can be used in higher percentages than dark malts before the flavors become clumsy and overbearing. Higher color varieties can (and do) overwhelm a beer when used in excess. It is especially important that the brewer taste caramel malts, as the flavor range varies more for a given color than with other malts. Substitution of like-color crystal malts can cause perceptible changes to finished beer to a degree not seen with ingredients like black malt.

Special Hybrid Malts

Color Range: 50–150 SRM

Special Hybrid malts are made by first caramelizing, then roasting, green malt. These have flavor notes that combine the attributes of both caramel and roasted malts. They can express as deep, dark, dried fruit aromas (raisin, plum) and are often used in rich dark Belgian style beers. Special B is a great example of this type and is integral to the Belgian Dubbel style.

Roasted Malts

Malts in this group are produced by loading pale malts into a drum roaster and roasting them, which destroys their enzymatic potential. They are heated to produce colors ranging from light brown to very dark. This group differs from caramel malts and tends to have more dry and

---

[*] A greater discussion of caramel malt can be found in the Specialty malts chapter.

astringent flavors. These malts need to be used sparingly, and it is rare to see recipe formulation that includes greater than 10 percent inclusion of any of the malts from this group.

Biscuit Malt

Color Range: 20–30 SRM

Biscuit malt is produced in the kiln at high temperatures; up to 440°F (227°C). The bread crust and toasted flavors that develop as a result are key components of nut brown ales. They tend to impart a dry finish and have characteristic nutty, toasted biscuit flavors. The flavors share some characteristics with Vienna malt but are much more intense, and the malt lacks enzymatic potential. Biscuit and Amber malts are fairly similar, but the prior has a slightly drier character than the latter.

Amber Malt

Color Range: 20–36 SRM

Amber malt is a lightly drum roasted malt that comes from the English malting tradition and tastes like toffee, baked bread, and nuts. The dry roasting process promotes the formation of pyrazine and pyrrole compounds that also impart some bitter characteristics. The flavor of amber malt works particularly well with cask conditioned mild ales as the dry finish provides a counterpoint to the estery aromas created by the yeast.

Brown Malt

Color Range: 40–150 SRM

Brown malt is similar to amber malt, but is given additional time to develop more flavor and color. Like Amber, brown malt has toffee, baked, and nutty flavors. Because the production of this malt promotes a different set of reaction pathways that are present in very dark malts, brown malt is sometimes used to add depth and complexity to darker beers. When used in excess, beers tend to finish with dry harsh notes that may parch a drinker rather than refresh them.

An early type of brown malt, known as blown or snap malt, was made using the intense heat by adding wood bundles to the fire. The resulting malt had a smoky flavor with caramel overtones. It was integral to the production of Porters prior to the development of drum roasters in the mid-1800s.

Chocolate Malt

Color Range: 350–500 SRM

Chocolate malt is a drum-roasted malt that provides dark color to beer. The mild burnt flavor it carries pairs well with the rich coffee and chocolate notes that are developed as Maillard products when it is roasted. It is not as dark as black malt, and has incredible depth of flavor with slight astringency. These malts tend more toward chocolate notes with less of the acrid flavors prevalent in very highly roasted malts like black malt.

Black Malt

Color Range: 435–550 SRM

Black malt provides a huge color addition to beer. The high temperatures that produce this malt cause the development of acrid flavor products that define stouts. These bitter, dry, and burnt flavors are somewhat moderated by the roasting process. Seeming to fly in the face of logic, highly colored black malts lose both flavor and color when they are roasted too much. As the temperature rises higher and higher, the flavor and color compounds basically turn to charcoal. The nearly incinerated grain mass will give little contribution to beer. It takes a skillful operation of the drum roaster to coax peak flavor and color out of the malt without completely charring it.

Debittered dark malt and barley have wonderful properties that allow a brewer to create a dark but crisp beer. Astringent flavors become quite pronounced as barley husk is roasted to very dark colors. By using huskless barley, the maltster is able to make dark malt with minimal bitterness. Although the dry and ashy aspects define hearty, full bore stouts, there is something magical about a beer that is optically dark but has bright, light flavors that seem to be in contrast with its visual appearance.

Roasted Barley

Color Range: 300–650 SRM

Unlike roasted malt, roasted barley is created before the barley is malted. The raw, dried barley can be roasted to a range of color roughly corresponding to the range between chocolate and black malts. Roasted barley is normally milder than roasted malts, but acrid, dry, and burnt flavors still dominate its profile. Black barley tends to produce a lighter colored head than corresponding colored malt does. This ingredient is a key flavor component of Dry Irish Stouts. Brynildson loves the cocoa-mocha flavors and aroma of 350°L roasted barley. He also likes the chocolate and roast flavors that are present in the lighter roasted products.

Special Process Malts

Special Process malts are made, not surprisingly, using special processes, which lead to a range of functional and flavor attributes unique to this group. Malts in this class include acidulated malt, smoked, and peated malts. The market for these malts is very limited, and they tend to be relatively expensive.

Acidulated Malt

Color Range: 2.2–4 SRM

Acidulated malt (Sauermalz) is made by promoting the growth of lactic acid bacteria during the germination stage by spraying sour wort on the malt prior to kilning. The resulting malt has a bracingly sharp flavor and reduces mash pH values while still complying with the Reinheitsgebot. This malt is best used as a light spice in a grist bill. At low usage rates, it is virtually unnoticeable but brings a bright acidity that can enhance an otherwise drab malt component. In most situations, if a brewer can taste traces of the acidulated malt, they've used too much.

This is a malt style that will see many variations as maltsters increasingly explore what unique flavors can be coaxed from barley and other grains. Many brewers are excited to see what sour, tart characteristics can be developed in beers when using acid malts. With an increased malt palette to experiment with, the development of unique and tasty sour flavors in beer can come from sources beyond long tank or barrel aging with traditional souring microflora.

Smoked Malt

Color Range: 2.5–5 SRM

Smoked malt is made by drying the malt wholly or partially using the direct combustion gases of a wood fire, which imparts an intense smoky flavor. Traditional versions (associated with the city of Bamberg, located in Franconia) use beechwood as the fuel source. Other special woods (such as cherry and alder) are used in other areas of the world. Lightly smoked versions of this malt can be utilized for up to 100 percent of the grist charge.

It is vital that smoked malt be used at appropriate levels; excessive use completely overwhelms underlying flavors and results in undrinkable, monochromatic beers. In general, US made smoked malt tends to be stronger than the European counterpart. The smoke flavor itself is divisive; many people simply do not like this flavor, but others think it gives unique character to certain styles of beer.

Peated Malt

Color Range: 1.7–2.5 SRM

Peated malt is a subtype of smoked malt that uses peat as the kilning fuel and flavor source. These malts are mainly made for the production of Scotch whiskey. They have a very strong and distinctive Band-Aid® like phenolic flavor that can easily dominate beer. Most veteran brewers agree that these malts are always best used in whiskey, as most beers made with peat are universally unpleasant.

Malts Utilizing Other Grains

This group includes malts made from grains other than barley. Because these other grains can be processed like barley malts, they contain a wide range of color and flavor. For example drum-roasted wheat malt can reach colors up to 550 SRM.

Wheat Malt

Color Range: 1.5–3.5 SRM

Wheat malt is, as the name suggests, malt made from wheat instead of barley. Because virtually any malt production technique can also be applied to wheat, products such as highly roasted wheat malts are available to adventurous brewers. Wheat has functional properties for beer as well; high protein content promotes foam formation and retention. Many wheat beers are not filtered, and some of the other properties that typify wheat beer styles (such as haze and a slightly bready flavor) may be ascribed as much to wheat as to yeast. Wheat is used in a variety of beers styles, such as Weizen and Berliner Weiss.

Wheat can be difficult to work with both in the malthouse and brewery. High protein and gum levels may require additional effort from an enzymatic perspective. Gibberellic acid use in the malthouse seems to be more common with wheat than barley. More intensive mashing regimes may be required to use wheat malts (especially those that are undermodified) in the brewhouse. In mashes that contain very high wheat malt percentages the use of rice hulls to create and maintain a porous mash bed structure can speed up an otherwise lengthy lautering phase.

Not much attention has been paid to the varieties of wheat used for making malts. Very few brewers seem to know what variety their wheat malt was made from. Wheat types are defined by protein level (hard/soft), color (red/white), and planting time (winter/summer). The red color of the wheat berries present in the bran of the grain comes from the greater levels of phenols and tannins contained therein. Harder wheat has higher protein levels. The size of the grains may differ; red wheat tends to be physically smaller and thus more difficult to mill consistently.

Rye Malt
Color Range: 2.8–3.7 SRM

Rye malt has some similarities with wheat malt but has a distinctive spicy flavor that translates well into beer. The combination of rye flavor with that of bold American hops marry well in beer, and judging from the number of beers that are constructed this way, many different brewers agree. Rye, like wheat and oats, lacks a husk, which contributes to the gummy, dense, and viscous properties manifested in both the germination bed and the lauter tun. As with wheat and barley, rye can be malted using a variety of techniques. Drum-roasted rye malts are available in colors up to 250 SRM (and presumably higher if desired).

Oat Malt
Color Range: 1.6–6.5 SRM

Oats are generally used as unmalted adjuncts. However, oats can be (and sometimes are) malted. Adding oats to a beer's grain bill typically gives it a full, soft, and silky mouthfeel. Malted versions of oats (like Simpsons Golden Naked Oats) have a rich granola like flavor that carries through in some beers. Bell's has made a number of one-off, small batch beers that feature malted oats to good effect. Although large amounts can make lautering a very annoying process, because of the high gum fractions, the flavors and texture that oats bring to finished beers are well worth the extra effort.

Distillers Malt

Color Range: 1.2–2 SRM

Distillers malt is generally made from a lower grade of barley than brewing malts, and typically not used by brewers. This malt has a higher husk fraction, smaller kernel size, and very grassy flavors. Production processes are tailored to maximize the available enzymatic potential, which allows distillers to use them at low levels in a cooked cereal mash. Over the days-long mash fermentation used by distillers, the enzymes from the malt break down adjunct starches to sugars. This malt does not add significant flavor to distilled spirits and much is lost during the distillation process. The use of distillers malt is not recommended for beer, as the flavor is generally harsh.

Chit Malt

Color Range: 1.2–2 SRM

Chit malt (also known as "short grown" malt), is made using a very short germination schedule. Grain modification tends to be very low, and although it shares many properties with unmalted adjuncts, it is allowed under the Reinheitsgebot. Chit malt is mainly used for head retention but it can be difficult to work with due to high beta glucans levels. For most brewers, the extra expense of a material that behaves like an unmalted cereal adjunct is not worth their time or energy.

Malt Flavor Descriptors

Malt and wort flavors span a large range of expression, and the subjective nature of flavor and aroma makes standardization of descriptors difficult. Many flavors that are present in beer can be attributed to a single compound; the buttery aroma of diacetyl, for example. Malt flavor is often very complex and results from the interactions of multiple molecules. Advanced flavor training benefits from having standards to work from, but many of the flavors associated with malt are not easily standardized. Two of the best efforts (that offer the most comprehensive flavor profiles) include the Malt Aroma Wheel® developed by Weyermann® and the Malt Tasting Glossary published in the *MBAA Technical Quarterly* by Murray in 1999.

Weyermann® Malt Aroma Wheel®
Weyermann CARAMUNICH®: Whole Kernel

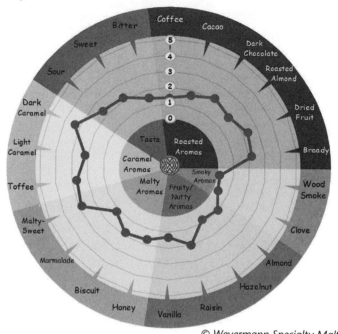

© Weyermann Specialty Malts

Fig. 7.2: Weyermann Malt Aroma Wheel® reproduced by permission.

Malt Flavor Descriptors (Murray, 1999)

CEREAL	Cookie, Biscuit, Bournvita, Cereal, Hay, Horlicks, Husky, Malt, Muesli, Ovaltine, Pastry, Rusks, Ryvita
SWEET	Honey, Sweet
BURNT	Burnt, Toast, Roast
NUTTY (GREEN)	Beany, Cauliflower, Grainy, Grassy, Green pea, Seaweed, Bean sprout
NUTTY (ROAST)	Chestnut, Peanut, Walnut, Brazil nut
SULPHURY	Cooked vegetable, DMS, Sulphidic, Sulphilic
HARSH	Acidic, Sour, Sharp
TOFFEE	Toffee, Vanilla
CARAMEL	Caramel, Cream Soda

COFFEE	Espresso Coffee
CHOCOLATE	Dark Chocolate
TREACLE	Treacle, Treacle toffee
SMOKY	Bonfire, Wood fire, Peaty, Wood ash
PHENOLIC	Spicy, Medicinal Herbal
FRUITY	Fruit Jam, Bananas, Citrus, Fruitcake
BITTER	Bitter, Quinine
ASTRINGENT	Astringent, Mouth puckering
OTHER	Cardboard, Earthy, Damp Paper
LINGER	Duration/Intensity of aftertaste

8

Barley Anatomy and Agriculture

Barley was first domesticated from a wild ancestor in the fertile crescent of the Middle East (*Hordeum vulgare spontaneum*) about 10,000 years ago, making it one of the oldest cultivated food crops. As human civilization spread, barley was successfully introduced into a diverse range of environments. Although not well suited for warm and humid climates, adaptation to extreme cold and elevated salinity have allowed the crop to successfully grow in areas ranging from subtropical to subarctic. Barley's environmental range is greater than any other cereal crop, and unlike some of its cousins, it is able to thrive in high altitude, arid conditions.

Although a comprehensive overview of barley physiology and agronomy is beyond the needs of most brewers, it is useful to have a cursory knowledge about the plant when discussing malt. At a biological level, the most basic function of a barley plant is to produce more barley plants, and it does so by making grain. This chapter will explore this natural cycle and its importance to malt and beer.

Plant Development and Structure

When a barley kernel germinates—in the malthouse or in the field—the first growth to emerge are the rootlets, otherwise known as chit. As the plant sprouts, this outshoot branches into individual roots. Another part of the kernel, the acrospire, originates in the embryo but grows underneath the husk material, eventually emerging at the distal end of the grain. The malting process halts growth at this point, but for a kernel planted in the ground, once the acrospire (stem shoot) reaches the soil surface, a first leaf emerges. The tubular stem continues to grow upward, while nodes periodically form and eventually grow into additional leaves. Each leaf has a sheath section that wraps around the stem as it extends upward before eventually branching away as a flat blade. Additional stems will emerge from lateral buds of the first stem. Depending on water and nutrient availability, barley will grow up to five "tillers" or stems.

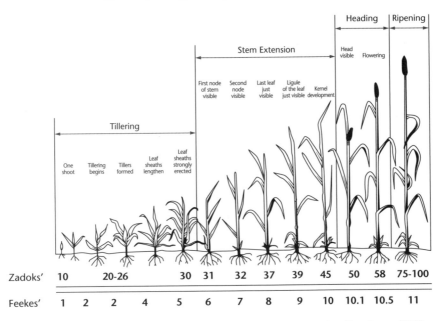

Fig. 8.1 Stages of barley plant growth from emergence to maturity. Courtesy of W.E. Thomason, et al. through Virginia Cooperative Extension, Virginia Tech and Virginia State University.

As with all grasses, the leaves of the plant alternate bilaterally in the direction of growth. The barley plant will develop about six nodes on each stem. A flowering seed head will eventually emerge above the top (or "flag") leaf. Not all tillers develop grain heads, but modern varieties tend to have a greater percentage that do, in part because yield is an important genetic selection criterion. As the plant continues to grow above ground, additional supporting roots grow below ground. By the time the plant reaches maturity, the total root system can reach as deep as six feet.

Barley plant growth can be grouped into three distinct phases; vegetative, reproductive, and grain filling.* Barley is mainly self-pollinating and because of this, varieties are relatively stable throughout successive generations. During the reproductive phase, the grain heads begin to form while still enclosed within the "boot" (flag leaf sheath). Barley kernels start as simple flowers called "spikelets" arranged along a central stalk called the "rachis." Six-row barley has groups of three spikelets that develop on alternating sides of the rachis. Two-row barley only has a central pair of spikelets that produce kernels. After the grain head emerges from the boot, the grain filling phase begins.[1]

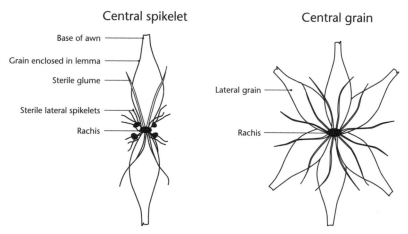

Central spikelet

Base of awn

Grain enclosed in lemma

Sterile glume

Sterile lateral spikelets

Rachis

Central grain

Lateral grain

Rachis

Fig. 8.2: Diagram showing the growth pattern of two-row and six-row barley kernels.

During the grain filling phase, kernels form, elongate, and then fill with the starchy endosperm and embryo material. Early in development, the endosperm resembles a milky fluid; as it matures, it becomes more like dough. When it reaches maturation, the grain dries up and shrinks, and the kernel becomes dense and hard. The overall form of the grain ear is dependent on kernel density. In varieties with ample spacing, the ear takes on a curved or "lax" form; tighter spacing produces straighter ears.

---

* The Zadoks Index further breaks down grain growth into nine defined stages; germination, seeding development, tillering, stem elongation, boot, head emergence, flowering, milk development in kernel, dough development in kernel, and ripening.

The Barley Kernel

Much like an egg, a barley kernel contains an embryo and a food source, both of which are enclosed within a protective shell. Each of these three elements has additional attributes that play important roles in malting and brewing. Figure 8.3 shows the point where the grain was attached to the rachis during growth. This is called the basal end, and is where the embryo is located. A small indentation, called a furrow, exists on the ventral side of the grain. The husk is composed of two distinct and over-lapping structures: the palea and lemma. The palea covers the ventral side while the lemma is located on the opposite, or dorsal, side of the grain. The awn, the characteristic long, rough bristle, is attached to the terminal end of the lemma. Although integral to photosynthesis during the plant's growth, the awn breaks away during harvest.

Close examination of a mature barley kernel reveals that while it looks relatively one-dimensional, it is actually composed of many distinct layers. Many consider the kernel a type of seed (especially from a botanical per-spective) but barley, like other grains, is technically a type of fruit called a caryopsis. In fruits such as peaches, the outer fruit wall or pericarp layer is thick, succulent, and edible. In grains, the pericarp is located directly beneath the husk and only a few cells thick. The next layer is the testa, the tough outer seed coat that protects the embryo and energy reserves. The dark outer coloring of a peach pit or an apple seed is also a seed coat. In barley, the testa and pericarp are fused together and act as a shield, preventing moisture and other environmental factors from reaching the living tissues they protect. Polyphenols (also known as tannins) are con-centrated in the testa layer. Barley husk is composed of tough and abrasive materials like lignin, pentosans, hemicelluloses, and silica, making it resistant to enzyme degradation during mashing and able to provide the stable structure needed to maintain a porous bed during lautering.

An important component of barley and wheat's chemical makeup is ferulic acid.[2] This organic compound concentrates in the cell walls of the seed coat and aleurone layer where it cross-links with arabinoxylan (a hemicellulose) to help hold the cell walls together. Ferulic Acid (3-methoxy-4-hydroxycin-namic) is the precursor used by yeast to produce of 4-vinyl-guaiacol (4VG), which is the clove-like aroma found in wheat beers.

Beneath the testa lies the aleurone layer. Like the pericarp, this special-ized layer is only two or three cells thick and is active in mature grain, unlike

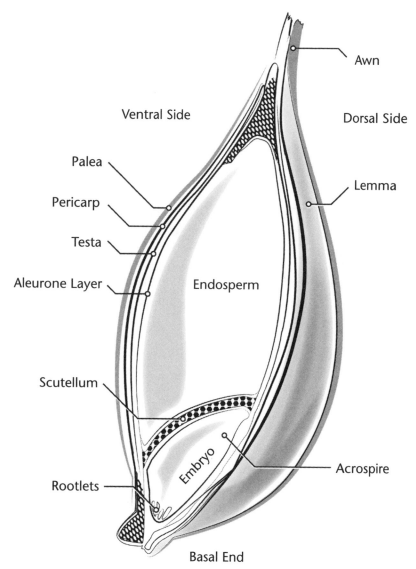

Fig. 8.3: Diagram of barley kernel components. Illustration: Alexander Smith

the enclosed starchy endosperm or the covering husk material. This layer does not grow but produces enzymes that access the stored energy reserves in the endosperm needed by the embryo for successful germination.

The starchy endosperm can be up to 80 percent of total grain weight, making it the largest structure in mature barley kernels. The internal structure of the endosperm consists of a mixture of large and small starch granules embedded within a protein matrix. The large starch granules are about 25µ (microns, or 1/1000th of a millimeter) in size and contain about 90 percent of the starch in the grain. The smaller granules measure roughly 5µ in diameter, and account for the other 10 percent.

The barley embryo rests at the proximal end of the barley kernel. In the quiescent state, the embryo is quite small; only about four percent of the total weight of the grain. During hydration and subsequent germination, the embryo enzymatically breaks down the reserves contained within the endosperm, using the energy to generate the new plant tissue.

Despite being self-pollinating, reproduction in barley requires two separate fertilizations; the first for the formation of the embryo and the second to give rise to the aleuronic layer and starchy endosperm. Interestingly, the cells are triploid, meaning they have three sets of seven chromosomes; two from the mother and one from the father. The cells of the embryo and the resultant barley plant are diploid, with a total of 14 chromosomes.

Barley Diseases

There are many natural threats to barley, wheat, and rye production, including diseases like mosaics, mottles, mildews and molds, bunts, blights and blotches, smuts, stripes, smudges, scabs and spots, rusts, and rots. All of these viral, bacterial, or fungal infections can wreak havoc on the development and viability of a barley plant.

Barley is particularly susceptible to infection during the early stages of plant development. As the spike slowly transforms into a mature kernel, disease agents can infiltrate the internal structures before the hull has fully developed. Growth on the nutrient-rich pericarp causes discoloration of the grain, "staining" it, and providing good visual evidence of barley growth conditions and disease presence.

A common fungal disease is "scab" or Fusarium Head Blight (FHB). In grain it will produce mycotoxin deoxynivalenol or DON.[3] DON carries the self-descriptive nickname of "vomitoxin" because of its harmful effect on animal digestive systems. Elevated DON levels in the grain have a few unacceptable implications specifically for the brewer, as beer brewed from grains carrying high DON levels is very susceptible to spontaneous gushing. Having a bottle violently discharge its contents all over the kitchen counter after removing the cap is not generally viewed positively by the beer-loving public. Attributing this activity to detectable levels of vomitoxin doesn't make for very good marketing, either.

Fig. 8.4: Effects of Fusarium Headblight on Wheat © Gary C. Bergstrom, Ph.D, Cornell University

FHB disease can be harbored by infected corn (maize) residue. Breeding advances have increased the agronomic performance of corn, and allowed it to be profitably grown in traditional small-grain growing areas, increasing the likelihood of FHB infection. Infected grain is generally smaller, shriveled, and may be discolored. The fungus stops growing when the grain is harvested and dried, but the toxin remains. DON levels in badly infected grains can exceed 20 parts per million (ppm). The FDA has determined that the DON level in food for both humans and swine should not exceed one ppm. The rinsing associated with the steeping step in malting can reduce the level of DON in affected grain. Barley carrying greater than one ppm DON, however, is generally rejected for malting use by most brewers and maltsters. DON is measured in grain via several laboratory methods, including Enzyme Linked Immunosorbent Assay (ELISA) and Gas Chromatography-Mass Spectrometry (GC-MS).

Barley and wheat breeders toil to increase FHB resistance in their crops due to the significant economic impact of the disease (topping

100 million dollars annually).[4] The US Wheat and Barley Scab Initiative (USWBSI) is a network of scientists, growers, food processors, and others who are jointly working on control measures to combat this widespread fungal problem.

Ergot

Ergot is a fungal disease that manifests when a spore successfully infects a grain flower. Barley is not particularly susceptible, but rye, with its open flowering, sees a high occurrence of infection. For the farmer, prevention is the best strategy as control is lost once it gets into the field. Tillage and crop rotation are good preventative practices. Ergot-infected grains produce a host of alkaloids that are biochemically active, and when ingested by humans or other animals cause the condition known as ergotism. "St. Anthony's Fire," as the disease was known in the Middle Ages, causes convulsive seizures and gangrene. Ergot also produces ample quantities of Lysergic Acid Derivatives (i.e., LSD), so infected grain can induce hallucinations and irrational behavior. The lush and fantastic visions that appear in some medieval European art (such as the works of Hieronymus Bosch) may have been partly influenced by this disease, given how common it was during the time period. The unseen fire which burned the limbs and minds of struck villagers was seen as the work of the divine, and some considered the disease's manifestations nefarious effects of witchcraft.

Rusts are another fungal group of plant pathogens that can affect barley, wheat, and other small grains. As they mature, stem rust fungi become visible as dusty, rust-colored masses and lesions on the stem and leaves of the plant. The fungus is windborne, spreads quickly, and reproduces rapidly, causing nearly complete destruction of an affected crop. The cycle of infection, incubation, and propagation for stem rust may be completed in as little as seven days. At the conclusion of the reproduction cycle, tens of thousands of spores are picked up by the wind. If conditions are right, they can be carried hundreds of miles before initiating a new cycle of the disease.

In 1916, nearly 300 million bushels of wheat were lost due to stem rust in North America. The last large-scale stem rust infection in the United States occurred in the 1950s and was responsible for a nearly 40 percent crop loss. Due to the devastating effects on agriculture, both the US and the USSR developed biological weapons based on stem rust during the cold war.[5]

"Today, wheat feeds more people than any other single food source on the globe," notes Norman Borlaug, the Nobel Prize winning author of *"Father of the Green Revolution,"* in the 2008 *New York Times* opinion piece, *"Stem Rust Never Sleeps."* His thoughtful analysis of the stem rust problem and the discussion of the possible eventualities of its spread called many scientists and botanists to action.

In the article, he notes that in 1999, a new strain of stem rust (*Puccinia graminis*) was discovered in Uganda. Although fundamentally similar to stem rusts currently found in North America, most wheat and barley varieties have no resistance to this strain (dubbed Ug99), and when infected, rapidly become unharvestable. Initially this strain was confined to a limited area of East Africa but has recently started to spread further. Development of barley varieties genetically resistant to Ug99 but also acceptable for brewing are still being developed, but it could be years before they see practical application across the country. Developing resistant barley varieties requires screening a wide range of plant material to identify resistant or immune genes.

If Ug99 established itself in the US, our first line of defense would be fungicides applied directly to the plant. There are several commercially available treatments such as *propiconazole* (marketed under a number of trade names such as "Tilt") that are quite effective against many fungi, including stem rust. Availability and supply would not present a problem as they are currently used on a wide variety of agriculture, but issues may arise for brewers; firstly, brewers are notorious for demanding minimal pesticide and fungicide on malting barley after heading (barleycorn development) begins. Secondly, residual fungicidal activity may negatively interact with brewing yeast. In addition to the possible effects on the finished beer, fungicides represent an extra cost that would almost certainly be passed on to the brewer.

Maturity, Lodging & Pre-Sprout

In order to prosper in the natural world, barley, like many other plants, developed a trait that prevents early germination. In the wild, after the seed matures and disperses, it remains dormant for a period of time. Only when the temperature and moisture conditions are right will it start to grow. Viewing the cycle from the perspective of the plant, this prevents the grain from starting its youthful and fragile growth right as the cold winter weather arrives. Barley and other plants that evolved to wait for the warm soils and ample moisture that signal spring had a distinct advantage over those that could not.

But this trait also creates a challenge for the maltster. How long post-harvest must she wait before the barley will properly germinate? If barley could break dormancy early the timing decision becomes trivial. So over the years, traits resulting in shorter dormancy were selected in breeding programs. This has resulted in some unintended consequences; if barley has no dormancy, then the mature seed, given the right warm and moist conditions, would start to grow immediately. Mike Turnwald, a farmer who works the Bell's Brewery barley farmland, once colorfully quipped that given the low dormancy of modern barley he would be nervous if someone spilled water in a growing barley field after mid-July.

"Pre-sprout damage" can be caused by a late season rainstorm. The falling rainwater soaks the plants, and the kernels in the seed head can actually start to grow while still on the stalk. Whether it starts in the malthouse or on the plant's stalk, germination triggers a cascade of enzymatic activity that profoundly changes the grain kernel. The increase in alpha-amylase activity due to pre-sprout damage will cause a breakdown of grain starch and the loss of potential extract. For wheat that is destined for baking applications, the destruction of starch is very problematic and therefore any affected grain is rejected or heavily discounted at purchase. The "Falling Number" viscosity test is a simple method of assessing the level of damage. Grain crops that suffer from pre-sprout damage are not affected homogeneously, and tend to have very inconsistent performance in the malthouse, which usually results in poor quality finished malt.

The Falling Number Test

The falling number test is a simple and practical method used to assess pre-sprout damage of grains. To perform the test, seven grams of grain milled to 0.8mm flour size is mixed with 25ml of distilled water. After the sample is mixed and heated for 60 seconds to make a gravy-like slurry, the time it takes for the weighted mixture to move through the sample is measured. The reported result includes the mixing time and is basically a measure of how viscous the sample is.

If the sample has had any germination activity in the field, then greater amounts of the enzymes responsible for starch breakdown change the consistency of the slurry. During the test, large starch structures are attacked by the enzymes and the resultant slurry has less structure and lower viscosity.

The test time includes the 60 second mixing phase and results vary from ~100 (the amount of weight that falls through slurry in 40 seconds) for grains suffering from significant pre-sprout condition, to over 300 for undamaged grains. The test is widely used as a basis for quality assessment all the way down to the grain elevator level.

Late season storms can have other consequences for the crop. A plant that has developed a top-heavy yield of grain can be easily knocked over by strong winds, heavy rain, and hail (also known and feared by farmers as the "Great White Combine"). A farmer must balance a grain's ability to remain standing and prevent "lodging" with potential yield. A variety that has a light head on top of sturdy straw would not be as susceptible to knockdown, but would probably not be commercially profitable, either. Breeding to prevent lodging resulted in the shorter variety called the "stander," because it remained standing when other varieties fell flat.

Variety Development

Since the 1950s, crops have dramatically increased in productivity. Known collectively as the "Green Revolution," these efforts have established measures of disease resistance to devastating plant pathogens through intensive, selective crossbreeding programs. The yield per acre has risen, allowing the world's population to soar without a significant increase of, or need for, cultivated lands.

Each year the American Malting Barley Association (AMBA) publishes a list of barley varieties acceptable for brewing. This short list is the culmination of years of work involving researchers, farmers, brewers, and Mother Nature. This process starts with the selection and crossbreeding of parent lines. Their offspring are assessed for many agronomic factors like flavor, yield, disease resistance, and more. Lines showing promise are regrown, and testing continues over a number of years in a variety of field conditions. If it performs well, this barley eventually reaches the malting trial stage, and finally the brewing trial stage. From first cross to initial commercial availability, named and accepted varieties can take 10 or more years of development.

Unlike corn (or humans), barley is mainly self-pollinating and therefore varieties remain genetically stable over many, many generations. Cross-pollination between two separate varieties is usually only accomplished by first removing the pollen-producing anthers from the barley flower before they mature to prevent self-pollination. A mature pollen donor is then used to create the cross. Harry Harlan's writings attest to the monotony of these manual techniques when done in large numbers.

After a cross is made, the genetic material of the offspring undergoes natural genetic recombination for a series of generations. During this time, desirable traits or specific genetic attributes can be accidentally unlinked and lost. This genetic drift occurs in barley for about three years. To battle this drift, geneticists have developed a process that uses double haploids. Immature pollen grains, which have only one set of chromosomes, (i.e., haploid) are developed into full plants using cellular cultures in specialized growth media. The single set of chromosomes is copied exactly in this process, resulting in a stable variety right out of the gate.

Modern biology and genetics have created an increasing number of technical methods to add traits to existing life forms, including the direct insertion of genetic sequences from other organisms. The resulting transgenic creations are called Genetically Modified Organisms (GMOs); philosophical and ethical debate rages about the merits of this approach. As it stands in the United States, there are no commercially available barley strains that are GMOs.[6] The research, development, and growth costs for a modified grain would exceed $100 million, and at this point there is no compelling reason for an investment of that size.

After breeding, growth, and harvest, the new barley seeds are planted for assessment. The progeny are evaluated for a variety of agronomic characteristics such as yield, straw strength, plant height, and kernel plumpness. Underperforming lines are culled and not regrown in subsequent years. Conventional disease screening is performed by exposing plants to pathogens and observing the results. "Marker Assisted Selection" is a newer genomic technique that is used to check plants for disease resistance. Quickly scanning the genetic code for pre-identified resistance genes allows for rapid culling of weak or susceptible plants. This powerful and rapid screening method does not involve genetic modification of the plant material, only identification of naturally selected genes. While significant advances in genetics have aided the process, plant breeding truly is a numbers game and a single named, accepted, and widely grown variety represents a tiny fraction of the crosses that the breeder began with.

As the observant reader could guess, the goals of barley variety improvement are to increase the positive and minimize the negative attributes. Some of the attributes that are of greatest interest to the breeders and growers include disease resistance, agronomic performance in varied conditions, enzyme and protein levels, and perhaps most importantly to the modern farmer, yield. What is desirable for one use may not be universally wanted everywhere. For example, the seemingly desirable trait of exceptional drought tolerance may raise protein content to a level unacceptable for certain brewers.

Taxonomically, barley is in the family *Poaceae*, the large group of grasses that also includes agriculturally important cereal plants such as rice (*Oryza sativa L.*), corn (*Zea mays*) and oats (*Avena sativa L.*).[*] Barley (*Hordeum vulgare L.*) has a closer relationship to wheat (*Triticum aestivum L.*) and rye (*Secale cereale L.*), as all reside within the same tribe[**] (*Triticeae*). Barley research benefits from the close relationship between it and the more agriculturally significant wheat crop. The $25 million federally funded Triticeae Coordinated Agricultural Project (T-CAP) supports vital breeding research for both barley and wheat.

Aegilops

Aegilops, the goatgrasses, are lesser-known but very important members of the *Triticeae* tribe. In most parts of the world they are considered weeds as they often compete with wheat crops and reduce farm yields. These small wild grasses were used as food in pre-agrarian societies along with wild einkorn (*Triticum uratu*) and barley. Derived from the German word for "one grain,"

---

[*] The word Sativa is the Latin adjective for cultivated and is used in reference to seed grown crops. It is related to sero, the Latin verb "to sow."

[**] Tribe is a genetic taxonomical classification between family and genus.

einkorn appeared about half a million years ago in the Fertile Crescent. Both einkorn and aegilops have seven pairs of chromosomes which encode about 30,000 genes. In contrast, humans have 23 pairs of chromosomes that in total have around 21,000 genes. These species are diploid; each parent contributes one chromosome to each of the seven pairs. At some point einkorn naturally crossed with a goatgrass (*A. speltoides*) to produce emmer (*T. dicoccoides*), which is a tetraploid species. This means there are four sets of each chromosome rather than two, and each parent contributes 14 pairs of chromosomes to the offspring. Domesticated emmer started to appear in archaeological records about 10,000 years ago. Further natural hybridization between emmer and a different goatgrass (*A. tauschii*) gave rise to the hexaploid spelt, the progenitor of modern bread wheat. The 42 chromosomes in wheat have 16 billion base pairs of DNA and about 90,000 genes, making it the most complex and largest known genome of any organism.

For some brewers, awareness of barley diversity is limited to the difference between two- or six-row types. Within the greater barley genus much greater variation exists than this simple distinction. As Harry Harlan traveled the world, he searched for, and successfully found, incredible barley diversity. Although the husk material is tightly attached in varieties used for brewing, some barley threshes free of the husk. These hull-less or naked grains are different than the pearled barley used in soup, which has had its husk material mechanically removed. The huskless varieties are grown as staple grains in China, Japan, Tibet, and Nepal.

In addition to international variants, other, somewhat minor differences exist within the same families of barley. Although brewing barleys have long awns that extend outward from the kernel, other types have extra flowers known as "hoods." The rachis, the central stalk that the barleycorns attach to, may be brittle or not. In some types of barley the aleurone layer has a blue tint. A single gene controls whether the barley plant produces two or six rows of kernels in the spike. Some varieties have been adapted to be planted successfully in the fall and over winter; others are planted only in the spring. Barley can even be annual or perennial. It is evident that modern malting barley is derived from a relatively narrow genetic pool when looked at in the greater context of the entire species.

Farming Barley

Commercial barley grows in many parts of the world, and is able to thrive in higher and dryer locales than any other cereal crop. While it can grow in warm or wet locations, barley prefers cooler and drier climates. Varieties can be found planted in arid North Africa, in the saline soils of Australia, on the high Tibetan plateau, as well as throughout Europe, Asia, and the Americas. Barley is not in widespread use as a food for direct human consumption, and much of it is used as livestock feed. For cooking, wheat has a preferred flavor and higher gluten content, which allows it to develop the structure needed to make bread and pasta.

The annual cycle of farming spring-planted barley in the US and Canada begins as soon as the ground is frost free and dry enough to allow access for farm equipment. Barley is planted one to two inches deep in rows six to eight inches (15–20cm) apart. The seeding rate is between 50–120 lbs/acre. Depending on temperature, the barley shoot will start to emerge from the soil three to ten days after planting.

Like many other crops, barley can benefit from fertilizer and micronutrients. Nitrogen promotes the growth of deep roots, and green robust plants. One bushel of harvested barley removes about one pound of nitrogen from the soil annually. Judicious use of nitrogen fertilizers

Fig. 8.5: The author at harvest time on the Bell's barley farm.

is essential to maximize yield and quality. However, it is widely accepted that excessive nitrogen can unacceptably elevate protein levels in the grain. Proper balancing of phosphorus and potassium is also needed to optimize yield.

Barley is grown under a wide range of conditions and susceptibility to disease varies by region. In areas with higher moisture and abundant corn, FHB may be the largest threat. Barley Yellow Dwarf Virus is passed by aphids and is more prevalent in winter barley. Winter barley is also more susceptible to powdery mildew. Stripe Rust sometimes strikes barley, but usually presents a bigger problem in wheat. In some growing areas, prophylactic fungicides are used to prevent common barley diseases, and weeds are controlled using herbicide application, but any preventative measures are done prior to heading.

Barley Economics & Cropland Competition

Barley acreage in the United States has seen a steady and significant decline since the 1930s. Part of the reason for this decline lies in the demands and pressures of two main markets: malting and animal feed.

Although growing barley for malting can be profitable for a farmer, it is also risky. Pay for malt quality barley depends on meeting quality specifications set by the maltster. Barley that does not make the grade must be sold to the lower-priced commodity feed market, sometimes at a price below the cost of production. There are ongoing efforts to develop federally supported crop insurance to protect against losses when the malting grade is not met.

Increasingly, barley is changing from a commodity crop to a specialty crop. To assure sufficient supply, maltsters or brewers will contract with a farmer to grow barley. Rejection or clauses for a discounted scale protect the maltster if predetermined quality specifications such as excessive protein content, low germination activity, or disease absence are not met. As interest in locally grown and malted barley increases, farmers and maltsters growing barley in nontraditional areas have discovered their own particular challenges. Andrea Stanley of Valley Malt in Massachusetts provides a local perspective, noting that, "Given our wet conditions, getting low protein is the least of our issues in New England. Our local barley has higher DON levels, pre-harvest sprout, and lower plumpness, bushel weight, and germination levels. So far we have steered away from discount conversations because we just need more barley planted, period. Beggars can't be choosers."

Farmland competition from other crops increases every year. Fifty years ago it was not economically viable to grow corn in the traditionally barley-rich areas of North Dakota and western Minnesota, and the costs for growing barley were certainly lower than those of corn. However, as livestock feed, all commodity crops compete on the relatively equal basis of nutritive value per dollar. The combination of high yielding, drought tolerant, genetically engineered corn, canola, and soybeans (as well as elevated crop commodity prices) has dramatically contributed to the substantial drop in overall US barley production.[7]

At the federal level, the "Freedom to Farm" bill, Conservation Reserve Program, and changes in federal crop insurance have also influenced barley planting decisions. In 1986, 608 million bushels of barley were harvested in the US. The average barley crop since 2010 has dropped to 192 million bushels.[8] In 1986, more than half of the barley grown was used for livestock feed and only about 20 percent went to malt. Today, the majority of barley grown in the US is destined for the malthouse. The previously large pool of barley that maltsters had been able to select from has shrunk considerably. This same trend is apparent in Canada, forcing maltsters not to be as choosy in poor harvest years.

The work of AMBA is focused on supporting systems to assure an adequate supply of high-quality malting barley. Since barley is a public-sector crop, it relies on adequate funding from a wide variety of stakeholders to remain current and viable. Brewers and beer industry members should be advocates of this organization and realize that the ongoing changes in the barley supply chain from evolving disease pressures to economic and climactic factors, make the efforts of this coordinating organization increasingly important for the world of US beer. Brewers and maltsters should educate themselves on the pressing issues surrounding modern barley production and get involved in the evolving conversations. The phrase "No Barley, No Beer" means that everyone, from large scale brewer to average beer enthusiast, has a stake in this issue.

References

1. P.M. Anderson, E.A. Oelke, and S.R. Simmons, *Growth and Development Guide for Spring Wheat.* (University of Minnesota Agricultural Extension, 1985), Folder AG-FO-2547.

2. Chris Colby, "German Wheat Beer: III (Mashing and the Ferulic Acid Rest)", *Beer and Wine Journal.* September 9, 2013. http://beerandwinejournal.com/german-wheat-beer-iii/.

3. *DON (Vomitoxin) Handbook*, United States Department of Agriculture Grain Inspection, Packers and Stockyards Administration Federal Grain Inspection Service. (Washington, DC., 2001), http://www.gipsa.usda.gov/GIPSA/webapp?area=home&subject=lr&topic=hb-don.

4. D. Demcey et al., "Economic Impacts of Fusarium head blight in Wheat", *Agricultural Economics Report* No. 396, (Fargo, ND: North Dakota State University, Department of Agricultural Economics, 1998).

5. F. William Engdahl, "Rust to Fertilize Food Price Surge", *Asia Times* April 4, 2008.

6. American Malting Barley Association (AMBA), *No Genetically Modified (GM) Varieties Approved for Commercial Production in North America,* (Milwaukee, WI, 2014), http://ambainc.org/content/58/gm-statement.

7. Karen Hertsgaard "Declining Barley Acreage", *MBAA Technical Quarterly*, Vol. 49, No. 1, (St. Paul, MN: MBAA, 2012), 25-27.

8. USDA National Agricultural Statistics Service http://quickstats.nass.usda.gov/results/71C4B26B-FFB1-3AF6-A8A6-A3835FDB8C22?pivot=short_desc.

Craft Micro-Maltsters

With the steady growth of homebrewing and craft brewing in the United States (or for that matter, the world), it really shouldn't be surprising that craft malting is growing as well. The science behind both malting and brewing is indeed complicated, but neither process is actually very hard to do; the operations and facilities of the micro-maltsters described in this section are living proof. The equipment and scale may differ from the big maltsters we have discussed in the previous tour stops, but the science and spirit are the same. Passionate and creative, the entrepreneurs responsible for these operations are building a network strongly reminiscent of the nascent craft brewing scene of 30 years ago. Small-scale craft maltings are springing up in varying geographies and climates like the United States, the United Kingdom, Australia, and Argentina.

Copper Fox Distillery and Malthouse
Sperryville, Virginia

The Copper Fox Distillery was founded in 2000, and in 2005, owner Rick Wasmund began a malting operation to produce all the malt used to make his unique whiskey. While touring the historic mill buildings that house Copper Fox, you get a sense of just how uncomplicated

◄ *Copper Fox Distillery in Sperryville, VA is housed in historic mill buildings.*

malting can be. Confident and calm, Rick explains his unified approach to making the malt needed for his fruitwood smoked and aged whiskeys. The process starts in Virginia's Northern Neck, where farmer Billy Dawson is contracted to grow Thoroughbred, the barley variety developed at Virginia Tech. Twice each week, 1200 pounds of that barley fills the steep tank back at the distillery. The grain is steeped for two days before being gently spread a few inches deep across an immaculate epoxy-coated concrete floor. The barley is manually turned periodically as it sprouts over the course of five days.

The green malt is transferred from the germination area to a room outfitted with a perforated floor located above another room equipped with a wood stove. As the fire is stoked, heat rises through the kilning bed and dries the malt over two days. Apple and cherry wood chips (added during kilning and aging) impart the flavors that play a starring role in the single malt whiskey. Chunks of locally grown orchard wood smolder atop the stove and fill the space with a pungent smoke that brings a unique quality to the malt. When the malt is finished, it is stored in large sacks until it is needed.

Touring the entire facility does not take much time; an expression of how simple the process can be. Malt analysis is an equally streamlined affair; hand and mouth are the primary assessment tools, and Rick generally uses the tried and true method of chewing malt to determine its quality.

The distillery proper is located at the other end of the barn. The malt is first processed using a hammer mill before being heated with water. The thin mash is moved into a tank where yeast is added and allowed to ferment for a few days before being moved to the still. The final, clear, 155 proof liquor is aged with more fruitwood chips before eventually being packaged. In addition to keeping his own distillery stocked, Rick also supplies malt for other distillers and a few brewers.

Valley Malt
Amherst, Massachusetts

The rich and productive soils of the Connecticut River Valley have long been used for specialty crops like tobacco. Andrea and Christian Stanley's Valley Malt Company is located in a barn on a semi-residential street in the farming town of Hadley, near Amherst. Barley production moved westward out of this area many, many years ago, but this operation supports the local farmers working to grow an unfamiliar crop.

Andrea sums up the approach, "Good malt starts in the field; you cannot make good malt from shitty barley. It needs to be well grown."

The couple's approach to making malt is a high-energy blend of education and passion. Christian is a mechanical engineer by training and trade, and Andrea has amassed a significant library related to the history and manufacture of malt. Prior to embarking on their venture, the couple received the Michael Jackson Fellowship, visited the eight small-scale floor malting operations in the United Kingdom, and took the four-day malting class at the North Dakota State University cereal labs.

In 2013, they plan to make about 75 tons of malt in their compact 600 square foot location using a pneumatic four-ton uni-malting system. Although computer-controlled blowers are used to maintain the desired temperature and moisture levels, the system is hardly mechanized; in fact, Andrea regularly pops inside the bin to manually turn the malt using a repurposed avalanche shovel.

The centerpiece of the compact and efficient design at Valley Malt is the bank of self-designed Uni vessels. The two 8 by 8 malting vessels are divided in half; each section is capable of holding one ton of grain. Barley is about two feet deep at the time of loading. A new batch is started every other day, and the entire lot of grain is submerged in about nine inches of 55-60°F (12-15°C) water to steep on a timer-based schedule. Andrea uses a pool skimmer to remove chaff and other unwanted floating material from the batch. There is a timed rest between each soaking step to allow the seeds the chance to uptake oxygen.

When it is time to move into the germination phase, blower and exhaust ducting are attached to the chamber. Both temperature and humidity are monitored and adjusted to maintain tight control of the germination conditions. The air may be heated or cooled, and additional moisture can be added via spray nozzles. The bed may reach 40 inches in depth by the end of germination.

Kilning, the final phase of the process, is accomplished by using a blower that is routed through a heat exchanger powered by a hot water boiler. During the first 18 to 24 hours, free moisture is removed from the grain using low temperature air. Curing (or kiln-off) starts after "breakthrough," when the malt becomes physically dry and the discharge humidity drops. Color and flavor continue to develop during this phase, up until the malt is cooled with room-temperature air.

Afterwards, it is sent through a debearding machine for deculming, a screen cleaner to separate undersized material, and finally to packaging and storage. The 40 to 50 customers are evenly divided between breweries and distilleries, in part due to a New York state initiative that lowers taxes for spirits made from at least 51 percent locally grown, in-state grain.

Because grain condition heavily influences malting performance, incoming barley is sent out for analysis. The falling number test data (see Chapter 8) is used to indicate any pre-sprout damage; extra care during the steeping air rest is needed to prevent the batch from becoming lactic.

In talking to Andrea and Christian, it's evident that their approach is heavily influenced by historic malting texts. Many of the quality assessments used in the past remain applicable for the craft maltster today; for example, the moisture content can be judged by simply shaking some grain in the hand to roughly measure density. A practiced and observant maltster can obtain valuable information about how a batch is progressing by simply smelling it: freshly germinating malt smells like cut cucumbers.

Powered in part by the 'locavore' movement, Valley takes special care to identify the farmer on the packaging, "People do want to connect with where their food comes from" says Andrea. Beyond local, there is a sense of adventure in attempting funky specialty mash ups such as cherrywood smoked triticale malt. Education and communication efforts range from Farmer Brewer Winter Weekend Conference to ongoing trials to locally grow over 60 historic land race barley varieties. They have had good results with Charles, Endeavor, Newdale, and Pinnacle varieties.

Michigan Malt Company
Shepherd, Michigan

Wendell Banks had a lot of commercial experience both as a brewer and as a certified organic farmer before starting a malting venture in a small farm town in central Michigan. He currently makes malt for several brewers and distillers across the state. Standing next to a pile of gently germinating grain, Banks patiently explains, "People have been making malt for 10,000 years; I do it like it was done for the first 9,500."

Wendell Banks stands outside his malthouse in Shepherd, MI.

The converted bean wagon at Michigan Malting. It can be used to steep, germinate, and kiln a batch of malt.

His floor malting process is distinctly different from that at Copper Fox. A specially modified stainless steel bean sprout drying wagon is the heart of the operation, and into this apparatus he loads 3000 pounds of cleaned barley or wheat to start a batch. After a two-day, intermittent soak, the now-hydrated grain remains on the grated false bottom as the water is drained away.

Initially Wendell would germinate inside the wagon, but as production increased, germination was moved to a flat concrete floor. As it grows, the grain is turned both manually and mechanically. The green malt is loaded 16 inches to 18 inches deep in the five-foot by ten-foot wagon for kilning. A stationary 1.4 million BTU blower/heater is coupled to the bottom of the apparatus, and heated air drives the moisture from the malt over the course of six to eight hours.

After drying, the malt is fed through an ancient Clipper 1B seed cleaner (to remove rootlets and small materials) before being packaged into bags. When malting losses are accounted for, the yield is about 2500 pounds of finished product. Wendell colorfully sums up his business model: "I used to make sprouts; I still do; now I just roast them at the end."

Colorado Malting Company
Alamosa, Colorado

The San Luis Valley sits high and dry at about 7500 feet above sea level. Annual precipitation rarely tops seven inches, most of which falls during the winter as snow. The sunny days and cool nights are perfect for cultivating barley, presuming the crop can be adequately irrigated. The Cody family farm has been growing barley for Coors for four generations. Jason and Josh Cody's great grandfather has a silver watch given to him for four decades of growing barley for Coors.

In 2008, Jason and Josh decided to take it a step further. They built a pilot malting facility and experimented with malting the barley themselves. The initial malthouse could malt 500 pounds of barley at a time. Initial skepticism from local brewers turned to loyalty as they began brewing with the Codys' malt. Increased demand led to expansion, and a new malthouse was subsequently built. Current capacity is about 500,000 pounds annually. In 2014, Colorado Malting Company (CMC) malted the 400,000 pounds of malt grown on their property and planned to purchase additional barley to produce more malt.

Rogue Malting
Tygh Valley, Oregon

Further west of Colorado, in Oregon's rural and arid Tygh Valley, Rogue Ales grows and malts both winter and spring barleys on its farm. Originally, the project was based at their Portland offices, but moving production to the farm has allowed the program to expand, and it now produces floor malts used in several of their specialty brands (like Good Chit Pils). Rogue runs a full analysis on every batch of malt. "Making malt is very difficult; it took us a year to figure it out," explains Rogue COO Mike Isaacson.

A 38-hour steep schedule is used to reach the desired 45 percent moisture content. It takes place inside of a repurposed open top fermenter that was retrofit with aeration nozzles and an overflow pipe. The concrete floor used for the four to five day germination phase has been epoxy coated to allow for better control of unwanted mold growth. The eight inch deep grain bed needs to be turned by hand multiple times each day; shoveling this particular pile only takes about 20 minutes.

The kiln is a four-foot by ten-foot perforated bottom box mounted about a foot above the floor. A burner/blower pushes air through the 18 inch grain bed from below. It takes about two days to bring the malt to 5 percent moisture content. The batch size is 1400 pounds. Rogue notes that they could do 2000 pound batches but "the kiln is just not as efficient at that scale."

Cleaning and deculming takes place on a screening table, and altogether it takes about 40 hours of work to produce 1200 pounds of finished malt. At 50 pounds of estate malt per hour of labor, this is clearly a labor of love for the brewery. Isaacson sums it up well, "The coolest thing is that your hands are in the grain; you are making malt."

A Whole World of Malting

While the vast majority of malt made worldwide is produced in highly automated, very efficient, large-scale malthouses, small-scale malting exists in countries other than America, too. After more than 30 years of producing malt on an industrial scale, maltster Grant Powell ventured out on his own. Powells Malt is located in south suburban Melbourne, Australia. Powell and his son Michael have been making Pilsner, Ale, Munich, and Wheat malt for the specialty market since 2003.

BA-Malt SA of Buenos Aires, Argentina, was founded by Martin Boan and his wife, Carolina in 2005. They make 1200 kg (~2650 lbs) of malt per batch, based on their kiln capacity and operations. Using the Scarlet variety barley grown in Argentina, they make a wide variety of base, high-kilned, caramel, and roasted malts. They also malt wheat and rye products. The craft brewing industry has been growing steadily in South America for many years, and BA-Malt currently sells malt to more than 300 microbreweries in Argentina, Brazil, Uruguay, and Paraguay. Martin reports that they have doubled their annual sales for the past three years in a row.

Despite being home to a rich brewing history, Scandinavia has been largely dependent on foreign supplied specialty malts for many years. However, the 2006 Nordic Brewer Symposium launched the development of small-scale malting businesses by the Oslo-based Nordic Innovation Centre. The resulting "Nordic Malt House" report provides the detailed breakout and analysis required to develop business plans for facilities large enough to produce 500 kg (~1100 lbs) batches of malt. The objectives section includes descriptors such as simple, flexible, modular, low tech, low investment, and small scale. The document is an invaluable source that enables local farmers and brewers to jointly own and operate a small malting house. Total investment for equipment sized to produce 300 tons per year is calculated to be less than $200,000 USD.

A Cautionary Note

While it may seem that all you need is a tank and a floor to start your own craft malthouse, the business of malting is really not that easy. In some regards it is like brewing: anyone can combine some malt, water, hops, and yeast and technically make a beer, but there is a big difference between the mere fermentation of a beverage and a product that people will actually pay good money for. The sales price needed for a boutique maltster to be sustainable (read: profitable) is significantly different than that for a highly automated major malthouse. Defining just what a small operation is able to deliver to the brewer or distiller is vital to the success of these businesses.

It is the same with malting, and the availability of malting-quality barley is less than you probably imagine. In fact, barley constitutes only 1 percent of planted acreage in the United States. In 2014, the harvest forecast for corn (maize) in the United States is 13.8 billion bushels, compared to 187 million bushels of barley—that's 70 times more corn! Barley production peaked in the 1980s, when roughly 30 percent of the total barley crop was used for malting. Most of the barley crop was used for animal feed, and maltsters could actually choose the cream of the crop for malting. Any barley variety can be used for animal feed, but some are not suitable for malting due to higher varietal protein levels and smaller kernel size. Malting barley needs to be low protein with plump kernels to give the best extract per pound to the maltster and brewer. Barley for animal feed on the other hand, should contain higher protein and the kernel size is immaterial, so obviously a big price difference exists between the two grades. The demand for malting barley has been fairly steady for the last 40 years, but overall barley acreage has declined; today malting accounts for 75 percent of the crop. This shift in supply and demand has changed the amount of risk a farmer assumes when planning his crops for the coming year.

A farmer needs to balance three factors when planning his crops: economics, agronomics, and labor. The economics are the prices he can get for crops in the coming year. While corn and soybeans have futures markets on the Chicago Mercantile Exchange, barley does not (being only 1 percent of the overall market). A barley farmer is not required to grow barley; if other crops look to be more profitable, then he (or she) may very well switch. In fact, a typical farm in North Dakota will be about 5000 acres in size, and of that land, 500-1000 acres may be barley (and probably more than one variety). This is where agronomics enters the picture; rotating different crops and different varieties of barley helps the farmer maintain the health of the land, stagger the harvest of each crop to better divide the amount of time and labor, and better manage the financial risk if any one crop should perform poorly.

Growing barley takes more experience and attention to detail than other crops like corn and soybeans. Farmers have a larger portfolio of crop protection products for corn and soybeans, including varietal breeding for disease resistance, pesticides, and herbicides. Farmers can largely just plant these crops, fertilize, and walk away until harvest time. A barley farmer, on the other hand, has to carefully gauge how much fertilizer to use to get a good yield, as too much fertilizer will increase the protein content (or more plainly, the total nitrogen). Barley is a cool-season grass, and the weather can have a significant effect on kernel development. Hot weather during pollination and early growth can kill the embryo and/or stunt the kernels, leading to small endosperms, and therefore higher relative protein levels, which lowers the price that a farmer can get for the crop.

Maltsters and brewers are very particular, and high protein in an otherwise perfect barley will still result is a price discount to the maltster. If the barley appearance and size is less than perfect, it will be dismissed as animal feed, which commands an even lower price. Why are maltsters and brewers so particular? Because any change in barley characteristics requires adjustment of standard procedures at the malthouse and brewery, and therefore more work. Time is money; it's as simple as that. The result of all these pressures means the barley market is becoming similar to the hop market, where contracts are the name of the game. There is some barley available for small maltsters on the open market but it is getting scarce. You may think that you can just call up your local farmer to have them grow the barley for you, but you have to understand the big picture and realize that she just may not be interested. Raising barley for malting makes more sense financially if the grower is also the end-user, as in the case of Rogue and Bell's breweries.

9

Barley Varieties

When asked to name a favorite hop variety at the start of a presentation for Kalamazoo Libation Organization of Brewers (KLOB), the local Kalamazoo brewing club, the home and professional brewers in the audience were able to quickly identify a long list of well-known hops. A lively discussion complete with head nodding, hand raising, and experience sharing ensued. A few minutes later, when asked to name a favorite barley variety, the list was limited to two barleys (Maris Otter and Golden Promise), a bunch of malt types (crystal, Munich, biscuit), and not much else from the noticeably quieter crowd. Barley isn't as sexy as hops, gets less attention in the beer community, and as a result, many brewers don't give it the attention it deserves. For many novice brewers, the "soul of beer" is defined by the processing of the malt during brewing, not by the attributes and quality of the product that journeys from the field through the malthouse to the brewery.

Dan Carey from the New Glarus Brewing Company scrutinizes barley quality attributes. The Swiss-themed village of New Glarus is located in the south central part of Wisconsin. The picturesque drive from Madison winds through bucolic wooded hills studded with family-owned dairy farms to a beautiful brewery overlooking the town from a hill to the south. Although its beers are only sold in Wisconsin, New Glarus Brewing Company is world renowned for their unique, high quality beer. Carey is a very comprehensive, practical, and well-educated brewer. A Siebel Institute classmate once remarked that, "If God had any questions about brewing, he should ask Dan."

When I talk to Dan, his insightful comments reveal a detailed yet holistic brewing philosophy. It is evident that he sees strong continuity from breeder to farmer to maltster to brewer to beer drinker. He feels that the key to understanding the malt is understanding the barley variety it's made from, where it was grown and where it was malted, and using that information to predict how the malt will work in his brewhouse and beer.

Dan justifies his philosophy with his perspective on the nature of beer. "I want to believe that the varietal characteristics are important. I want to brew in a world where terroir is important. So if someone like Eric Toft [of Private Landbrauerei Schönram, located in the Southeastern corner of Bavaria] puts on his label that these are hops from this farm, grown in this year, that is the world I want to live in because that's more interesting than just a commodity."

Dan is not alone in this thinking. Although many different varieties of barley malt have been successfully used throughout the history of brewing, some varieties function better than others in certain applications. Understanding and specifying particular barley varieties provides an additional level of control in brewing. As many brewers have likely experienced, all two-row pale malts are not created equal. To better understand how varieties differ from one another it is useful to know where and how they are developed.

Landrace Barley Strains

Not all barley is from a pure variety. Landrace barleys are not single, pure cultivars, but varietals that have changed slowly over time and adapted to their local growing conditions. Over the centuries, evolutionary competition selected those populations best adapted to survive and thrive during the cycle of planting, growth, and harvest. Since farmers retained their own seeds from the prior year's harvest, grains were (for the most part) very local. Agricultural historians believe that the barley grown across Europe in the early Middle Ages was mainly six-row. The genetic ancestors of two-row types grown in Europe were likely brought back from the Middle East during The Crusades.[1]

The landrace varieties most relevant today include the Bere (or Bygge, which was first found in Scotland), Hanna (which originated in what is today the Czech Republic), Oderbrucker (identified near Berlin, Germany), and Manchurian (an Eastern Asian variety also known as "Manshury").

Some successful strains only propagated through human intervention mixed with a bit of luck. Paul Schwarz, the well-respected professor and plant breeder from North Dakota State University, noted that before 1886, 80 to 90 percent of the barley grown in England[2] was a narrow eared, two rowed variety named "Chevallier." The eponymous Reverend John Chevalier described the origin of this variety: "A laborer [named Andrews], lived in a cottage of mine at Debenham in this county [Suffolk]. As he passed through a field of barley, he plucked a few ears, and on his arrival home threw them for his fowls into his garden, and in due time a few of the grains arrived at maturity, and as the ears appeared remarkably fine, I determined to try the experiment of cultivating them."[3] A subsequent variety, Goldthorpe, is named for the town in Yorkshire where a man named Mr. Dyson plucked it from a field of Chevallier, much like Andrews had for the original strain.

Barley Immigration

As the 19th century drew to a close, most barley varieties were semi-homogenous collections transported to new territories by emigrating farmers. The proliferation of the Manchurian type is a good example of this human-powered spread, "In 1861, Dr. Herman Grunow of Miflin, WI, obtained a sample of a promising six-row barley when traveling in Germany. He was told that a German traveler had found this 'Manshury' barley in 1850 along the Amur River, on the border between Russia and Manchuria."[4] This world traveling variety is the progenitor of all modern US six-row types.

Immigrants to North America brought prized barley seeds with them. In addition to Manshury (favored by German transplants), there were three other significant varietal lines that dominated production prior to 1900. The first two were the Scottish Bygge and English two-row, grown by the settlers of New England and New York. While these strains began to establish a foothold in the new world, they didn't gain popularity right away. Prior to the Civil War, beer did not have significant "share of throat," and cider was far and away the most popular alcoholic beverage in the country. Because of this, the malting barley market was not particularly large and most of the crop was used as animal feed.

Another type of barley arrived with early Spanish settlers. By the 1900s, a substantial volume of "Bay" or "California Coastal" barley (a variety that originated in North Africa, and found its way to California via Mexico) was growing on the west coast. Close examination of the adobe brick used in the 1701 construction of the Mission San Cayetano de Tumacácori in southern Arizona reveals traces of California Coast type kernels, suggesting that the grains arrived with the settling missionaries. In the late 1800s English maltsters and brewers had discovered the value of this robust six-row variety, and began to regularly import it from mainland Europe.

By the 1940s, two-row barleys made up only a small proportion of the US and Canadian crop because of brewer preference for six-row types, but limited amounts of two-row Hannchen and Compana-Smyrna types (of European and Turkish origin respectively) were grown in the Pacific Northwest. In addition to the more dominant varieties, six-row Tennessee Winter Group barleys (that probably originated in the Balkans) were grown in areas where the winter barley was able to survive. From the direct experience with high winter kill rates due to harsh weather that the Bell's farm experiences in Michigan, it is likely the varieties fared better in Tennessee. These barley types have largely been supplanted by corn for animal feed, but given the name and growing area, it is possible that they were a component of the "Kentucky Common Beer" that is described in older US brewing books.

The turn of the century heralded new interest in grain plant breeding and crop development. In the US, increasing demand for beer and a population that continued to move westward fueled and influenced plant breeders. By the mid-1900s many commercial barley varieties were being grown across the country. Historic publications describing these varieties give a voyeuristic peak into how brewers used these older strains. Modern brewers and beer drinkers know what Cascade, Liberty, Warrior, and Glacier hops taste like, but what about the barley varieties with the same names? Some type names seem to indicate the location they originated or thrived; Missouri Early Beardless, Michigan Winter, Ohio 1, Siberian, New Mexico, Dayton, Texan, Cordova, Nepal, and Soda Springs Smyrna. Others like Velvon 17, Wocus, Tregal, Dicktoo, and Wong are less obvious, but sound just as intriguing. What adventurous brewer wouldn't want to run a trial brew with the likes of Olli, Alpine, Sunrise, Winter Club, or Harlan?

By the 1960s, farmers were growing varieties specifically bred for malting qualities; prior to this, all barley was used indiscriminately as animal feed or for brewing, and the variety chosen by brewers was most likely whatever was available. In the years since, scientific advances in barley and industrial brewing's preference for two-row types have driven changes in the mixture of varieties. The 1960s produced several six-row types, including Wisconsin 38, Larker, Morex, Robust, Lacy, and Tradition. Around the same time, two-row types gained greater acceptance by brewers who now found that it could be used effectively to make their beers, and thus was more widely grown in the western US. Some important varieties from this time period (like Hannchen, Firlbeck, Piroline, Betzes, and Klages) have largely disappeared from commercial production due to the emergence of higher yielding, more disease resistant types.

Prior to the release of Klages in 1972, virtually all two-row barley grown in North America was developed in Europe; US and Canadian breeders mostly focused on six-row types. For the two-row Hanna, "the first record of its introduction by the USDA shows that it was received in 1901 from Kwassitz, Moravia, Austria."[5] The North American variety of Hanna is a composite of the natural and intentional selections that occurred in Hannchen fields of Northern California. Betzes was developed in Germany, and arrived in the US via Poland in 1938. The Moravian strain was brought to the US before and after WWII from Moravia (Czechoslovakia) by the Coors Brewing Company.

US and European Varieties

The way new varieties are developed for commercial production will be explored a little later in this chapter, but generally, as breeders identify useful traits (such as resistance to disease or increased yield), new types come into favor and replace older proven lines. Although varieties in the US were not specifically intended for brewing use up until the mid-1900s, development of new barley varieties today is driven by specific brewing attributes. Having knowledge of varieties is important because their differences affect malting, wort production, and ultimately the beer. Some of the varieties available today have been around for many decades, while others were developed much more recently.

Although the barley variety used to make malt is not always evident, increased brewer interest will lead to increased transparency. If purchasers don't ask about the barley that made the malt, there is no compelling reason for malt salespeople to provide the information freely. It is always in the best interest of the grower, maltsters, and brewer to ask if additional information on varieties is available as it helps to facilitate dialogue throughout the supply chain. As older varieties are phased out and newer ones supplant them, it is clearly in the interest of everyone to know and guide what comes next. An example might include resistance to disease or drought; if the flavor and performance of a new type is similar, but it performs better, then an eventual transition to the new type would be in the best interest of everyone involved.

The American Malting Barley Association (AMBA) annually publishes a list of recommended varieties for farmers to grow. The types on this list have demonstrated good agronomic and malting performance, and an AMBA brewery member must feel that it has acceptable flavor for brewing use. A brewer looking to learn more about barley varieties grown in the UK should refer to the Scottish Barley Variety Database.*

Variety Development and Acceptance

New barley varieties start in the hands of a plant breeder. To create a genetic cross, a breeder will manually pollinate the barley flower. The grain kernels of the pollinated plant will have a mixture of attributes from both parents. This is a numbers game; a game of genetic roulette in which the vast majority of the progeny have traits that are no better that those of the parents. A breeder may be looking for specific attributes such as resistance to disease, better tolerance to drought, higher yield, or less susceptibility to problems like lodging (grain falling flat in the field), but doesn't always find them when plants are crossed, even if their parents had said traits.

As the offspring grow, they are assessed for the desired traits and any unintentional defects, and are quickly removed if they do not meet the program goals. The crosses that pass the initial assessments are regrown over subsequent years for further testing (and possible removal if they show defects down the line). The breeder focuses on the agronomic performance of the plant. If a promising line is identified, it may eventually be tested for malting performance. The new line is tested for factors like even modification, good enzyme levels, and the ability to break dormancy quickly. It is only after agronomic and malting performance criteria are met, that flavor and brewing performance

---

* http://barley.agricrops.org/menu.php?

are considered. If the variety meets all the breeding goals, it becomes a "recommended variety" and is grown commercially. The entire process, from first cross to commercial acceptance, can take upwards of 10 years.

Commercial recommendation varies by country. In the US, AMBA coordinates this process. Maltsters and brewers jointly assess promising lines, and if a good candidate appears, an appropriate amount of barley is malted to test brewhouse performance and flavor on a commercial scale. If an AMBA member determines that the barley meets necessary quality standards, the variety can officially be added to the recommended list.

Many countries, including the UK, Germany, France, Canada, and Australia have similar systems in place to recommend new varieties, but each is slightly different. In the US most barley breeding is done via "public" programs, where breeding is done at public universities and supported by the USDA. In Europe, private seed breeding companies dominate barley and wheat development, supported by royalty payments from successful launches.

Winter Barley

Barley can be classified by when it grows; winter types seed in the fall and survive the winter by going dormant. Spring barley, as the name suggests, is planted in the spring. There is considerable interest in the development of more winter barley varieties as these types are generally higher yielding. True winter cereal grains such as barley, wheat, and triticale require a period of cold before they flower. Spring types do not require the cold, but some "facultative" types have adapted to survive through periods of cold and may be planted in the winter or spring. In addition to yield increases of up to 20 percent over spring types, winters have other beneficial attributes. They require less water and help to reduce erosion by stabilizing the soil over the harsh winter months. In certain areas, spring planting must be delayed due to wet conditions, while winter types are already in the soil and growing. This allows for an earlier harvest period, giving the farmer the opportunity to plant a second crop in the same year. Grain disease pressure tends to increase throughout the summer, and earlier maturing winter grain minimizes these problems. The likelihood of thunderstorms also increases as the summer growing season stretches into late July and early August. Strong winds plus the possibility of "pre-sprout" damage from rain usually equals many restless nights for farmers and brewers.

Dan Carey's Thoughts on Variety Difference

"We live in a very small world with a global economy. You may end up with very fine European malt made with Argentinean barley. In Germany, the institutes (VLB & Weihenstephan) approve new varieties; these barleys are bred by private companies. Barley is an important crop and it's a profitable for these companies to breed them. The range of requirements is very small. They may have moderate proteins (say eleven-ish), and S/T of 42–43. In America, with AMBA, it is very democratic. You have syrup brewers, adjunct brewers, and all-malt brewers; you have three different groups all requiring different varieties and if anybody wants (a variety) then it is approved. It is a much wider range."

"The variety Klages had an S/T in the low 40s. Today we are marching towards 50 percent S/T with higher FAN levels. It is a slow, deliberate progression towards increased modification and going farther and farther from our (craft) needs. The difference between all the malts we use today is very subtle. There is a slow march from Klages to Conlon; that is a fairly significant change. The six-row and two-row are starting to merge; a two-row grown on dryland under stressed condition is not very different from a six-row."

"Although I seem to be saying that today's varieties are fairly similar, I have not lost my religion. If you define the variety, where it is grown, in what year, and where it is malted, you know a hell of a lot more about the product than you do (by just) looking at a COA. If you flip it and you say 'I don't care what the variety is, I just want you to hit these specifications' that is being a 'bleistift' (pencil) brewer. I don't think the COA really describes what is really going to happen in mash. So, (the decision) to malt Harrington or Conlon it will make a difference."

"To me, European malt is fuller and richer in flavor. It can be huskier, but in general it is richer and it lends itself to higher hopping rates. People say that's an enzyme issue; it's higher residual extract, lower attenuation, but beyond the attenuation there is a fullness, a richness, a maltiness that we don't find in American malt. I don't think it's due to kilning practices, I don't think it's due to malting practices, I think it's due to varietal differences."

"I'm not an expert on English malt, but the whole concept that people point to Maris Otter and say that the reason it is so rich is because it has gone down to 3 percent moisture and it is 3.5 Lovibond (is probably not the case); it's a pale ale malt. I would say, let's send Optic and Maris Otter to a traditional floor malthouse and I would be willing to bet that you could (still) taste a difference. And if I was wrong that would mean that there are a lot of really stupid businesspeople that run these small breweries because everybody is paying a huge premium for Maris Otter; if it was simply the malting process then all the maltsters will be putting in floor malt houses and running the cheapest barley through them. It is clear to me that Maris Otter is different than Tipple or whatever new variety they are growing."

In the list that follows, the year of genetic selection or acceptance is given for many barley varieties. In general, more modern varieties have greater yields, resistance to disease, and other attributes that lead to their acceptance. Variety selection is not static, but changes from year to year. Farmers are interested in yields and achieving selection for malting. If the brewer wants a lower yielding variety, then they will need to communicate that desire and pay a premium for it.

Heritage Barley
Chevallier
Chevallier dominated English production throughout the 1800s. As noted earlier in the chapter, this barley was selected from existing stock in Suffolk, England. In 2012, Chevallier was revived by Dr. Chris Ridout of the John Innes Centre in Norwich, England, where it was grown, malted, and brewed with to reevaluate flavor and disease resistance.[6] Chevallier is the great-grandfather of modern British barley varieties and as such, its genetic stock remains strongly represented today. Modern brewers might find it intriguing to brew with this historic variety to gain a deeper understanding of what kind of challenges brewers of the period faced, and to find out what beer made from ancient strains of grain actually tasted like.

Golden Promise
Golden Promise was bred from the variety Maythorpe in 1956. It was added to the English recommended variety list in 1967 and was widely used in brewing and distilling throughout the 1970s and 1980s. It is a

semi-dwarf (short straw) two-row variety that performs well both in the field and in the malthouse.[7] Although other lines of this barley have been developed, the original line is still grown as premium specialty barley, mainly used for distilling.

Maris Otter

Maris Otter is beloved by many brewers for multiple reasons. Years ago it was a widely-grown, but as newer barleys demonstrated superior agronomics (easier to grow, higher yield per acre), production sharply declined. In 2002, Robin Appell Ltd., a grain trading company (and owner of Warminster Maltings), acquired the rights to the variety. Since then, this classic barley has staged a comeback, especially among brewers of traditional English cask ales. In addition to the perception of superior flavor,[*] brewers feel that the Maris Otter releases its extract into wort more easily during lautering. Farmers demand a premium for the grain because it has lower yield than other, more modern types, which ultimately means a higher malt price.

Modern Two-Row Barley Grown in North America
AC Metcalfe

AC Metcalfe was bred by the Manitoba station of Agriculture Canada (AC) in 1986, and entered Canadian commercial production in 1997. It was accepted in the US in 2005. AC Metcalfe is well suited to western Canadian and US production. It yields about 10 percent more than Harrington, and outperforms it in other important agronomic categories such as disease resistance.

CDC Meredith

CDC Meredith is a Canadian two-row variety developed by the Crop Development Centre at the University of Saskatchewan in 2008. It is a modern, high-performing variety that has been optimized for yield, disease resistance, and enzymatic power. It is well suited for use in high-adjunct brews such as American light lager.

---

[*] There is no universal agreement on the superiority of Maris Otter, but speaking from personal experience I have had some great malt made from this variety. Then again, no variety guarantees quality; especially in a bad crop year or with poor malting practice.

Charles

Charles is a winter barley variety released in 2005 by the USDA-ARS, Aberdeen, ID and the University of Idaho Agricultural Experiment Station (AES).* Although it has seen some success in western states, winter survivability remains problematic in other areas. The promise of winter barley in the form of better yield, agronomics and water usage efficiency is promising but the barley needs to deliver; Charles is struggling to do that in areas with inclement and damp winter weather. Every year Bell's has tried to grow it, they experienced huge problems with winter kill on the farm in mid-Michigan. It does very well in other locations but appears to not be suited well to the northern US climate.

Fig. 9.1: The very impressive and spotlessly clean Busch Agricultural Resources Malt House, located in Idaho Falls, ID. Annual malt production capacity of the facility is 310,000 Metric Tons (683 million lbs).

Conlon

Conlon was created by the North Dakota Agricultural Extension Service in 1988 and made the AMBA list in 2000. This variety grows best in western North Dakota and Montana. Conlon is a modern North American malting barley with agronomic and brewing performance traits that are particularly well-suited for brewers that desire high diastatic power and FAN; this is mainly larger adjunct brewers.

---

* http://www.cals.uidaho.edu/edcomm/pdf/CIS/CIS1166.pdf

Conrad

Released in 2005, Conrad is a product of Busch Agricultural Resources, LLC, a division of Anheuser Busch. It originated in Fort Collins, Colorado, and was test grown in Idaho Falls. It performs well under irrigated conditions in western states. Conrad is very high-yielding barley.

Expedition

Expedition is a new two-row spring variety released by MaltEurop for western US production regions. Recommended by AMBA in 2013, Expedition has excellent yield potential with a malt quality profile reportedly well-suited for the production of all-malt, low protein beers.

Full Pint

Full Pint is a relatively new barley variety, bred by Pat Hayes at Oregon State. A spring two-row, semi-dwarf barley with good resistance to disease that performs best in the Pacific Northwest, Full Pint was developed specifically with the craft brewing market in mind. Craft brewers express an interest in barley with both great flavor and solid brewhouse performance, suitable for use in all-malt brewing. Initially, Full Pint did not make the AMBA Recommended Variety list as it didn't appear to have many advantages in agronomics over existing cultivars, but interest from small brewers has given it some exposure in the market. Tim Mathews of Oskar Blues has brewed with it and likes what he has seen thus far. Full Pint is used in a few of their beers including the collaboration beer they did as part of Beer Camp across America with Sierra Nevada. Although the protein levels are relatively high, he feels that the malt has both good flavor and it imparts a great body to the beer.

Harrington

Harrington was bred from Klages, Betzes, and Centennial parentage in 1972 at the University of Saskatchewan. It was released to growers in Canada in 1981, and made the AMBA list in 1989. Although it is still grown as malting barley, acreage is declining due to conversion to higher yielding varieties. It is still considered a "malting standard" against which other barley is judged. The slow and incremental shift toward higher yielding, more disease resistant and faster malting barley types is difficult to see as one variety shifts to another but reviewed across multiple years, the changes are noticeable. The breeding efforts trialed today clearly diverge from Harrington and researchers have been able to deliver results.

Moravian 37 & 69

Moravian 37 & 69 are both products of the Coors breeding program that gained AMBA acceptance in 2000. Both are grown by contracted farmers in the Intermountain West for direct use by MillerCoors. Moravian (as the name suggests) originated in Europe and has been further developed in the US. Coors started this breeding program to find a bright, good barley suitable for growing in the states close to Colorado.

Morex

Morex (short for "more extract"), is a six-row type released by the Minnesota Agriculture Experiment Station in 1978. Like Larker (which came before it), and Robust (which was grown years later), Morex was a dominant variety for many years. Agronomically, Morex and its family are outperformed by modern six-row varieties such as Lacy (2000), and Tradition (2004), and has fallen out of production.

Pinnacle

Pinnacle was released by North Dakota State University (NDSU) and the United States Department of Agriculture - Agriculture Research Service (USDA-ARS) in 2007. It has consistently large kernel size and yield potential. Most of the barley bred and released about the same time as Pinnacle tends to have higher protein levels. Pinnacle is therefore of interest to some brewers who are looking for lower levels.

Thoroughbred

Thoroughbred is a six-row winter barley that was developed by the Virginia Agricultural Experimental Station (VAES) in 2012. It is generally grown in the Tidewater region of Mid-Atlantic States. It is the preferred variety used for malting at Copper Fox in Virginia, and some of the start-up craft malting operations in this area have reported several positive brewing attributes associated with this variety.

European Varieties

The growing conditions and purchasing expectations in the UK and continental Europe are different than those in the US and Canada. European maltsters and brewers demand both lower protein levels and lower enzyme values than their North American counterparts, mainly because the beers brewed in Europe are predominately all-malt.

Traditional barleys that have been grown in Europe do not necessarily grow well in North America. European brewers want lower FAN and high extract malt to brew their beer with, while the American malt economy is driven by efficiency. Barley breeding is mainly done by private companies, and the marketplace is regularly hit with new varieties boasting increased agronomic performance. Some European brewers voiced concerns that these breeding changes are made with minimal concern for ultimate flavor of the beer. Some European brewers will pay a premium to specifically order older varieties that they feel provide superior flavor.

In the UK, the barley and malt industry supplies not only breweries, but also the world-famous Scottish whiskey distilleries. Malts destined for spirits have different functional and flavor requirements than those used for beer. There is significant variety crossover between growers from the UK, Scandinavia, Germany, and France, and new varieties are marketed and grown across the entire European continent. In the UK, Concerto, (released in 2008 by Limagrain), currently has the largest market share. NFC Tipple (2004) and Quench (2006) were both bred by Syngenta; Propino (2009) is a subsequent cross of these two successful varieties that continues to grow in popularity throughout England and Scotland. Optic (1995) is an older line originally bred by Syngenta. At one point it held a 75 percent share of English malting barley, and is still grown by Sheena Kopman on her family farm in Aberdeenshire, Scotland. Dan Kopman, of the Schlafly Beer/Saint Louis Brewery, goes back to the family farm each year to purchase barley.

Older varieties that have seen greater adoption in continental production, and thus are more likely to be present in German malts, include Steffi (1989) which was bred by Ackermann Saatzucht, Scarlet (1995), and Barker (1996) both of which were bred by Josef Breun. These varieties often come up when speaking to flavor-focused German Brewers. Eric Toft, the American-born, Wyoming-raised brewmaster of Private

Landbrauerei Schönram in rural South-East Bavaria chooses Steffi for his beers. According to him, although it has a higher cost and is only about one percent of German production, it gives his beers superior qualities in flavor and foam. More modern varieties grown in Germany include Grace (2008), Marthe (2005), Propino (2009), and Quench (2006). Wintmalt, a winter barley that has shown good performance in Europe and is being trial grown in the United States, was developed by KWS Lochow in 2007.

For brewers, knowing the variety of barley the malt is derived from is important to understanding malt performance and specifications. Very few (if any) brewers simply purchase "aroma hops" without naming a variety, so it is surprising that so many turn around and simply purchase "two-row barley malt" without that same information. Given the complexity and impact on the final beer, barley and malt deserve the same attention as hops, if not more, when brewing certain malt-heavy styles.

Selection of specific barley varieties comes down to brewer preference. The variety influences how the malt is processed in the malthouse and how it performs in the brewhouse. Are there major flavor differences between varieties? Some brewers say yes, others say no. But variety has an undeniable and significant impact on the malting process, which ultimately manifests itself in beer. Even if it seems like a lot of work over a minor difference, a smart brewer will investigate the source of her malt, choosing a variety that best benefits her specific brewing or stylistic goals. To do that requires ongoing education and dialogue on the part of the brewer so that they can choose their malt wisely.

As Dan Carey puts it, "Knowing the barley variety helps me understand what the outcome is going to be. I hope that we are going to start to see more appreciation of the variety characteristics in modern brewing."

References

1.   Paul Schwarz, Personal Conversation with Author, 2014.

2.   E. S. Beaven, *Barley, Fifty Years of Observation and Experiment,* Foreword by Viscount Bledisloe, (London: Duckworth, 1947) 90.

3.   Walter John Sykes and Arthur L. LING, *The Principles and Practice of Brewing (Third Edition)* (London: Charles Griffin & Co., 1907), 421.

4. Paul Schwarz, Scott Heisel & Richard Horsley, "History of Malting Barley in the United States, 1600 – Present", *MBAA Technical Quarterly,* vol. 49, no.3. (St. Paul, MN: MBAA, 2012) 106.

5. Martyn Cornell, "Revival of ancient barley variety thrills fans of old beer styles." *Zythophile (blog)*, April 15, 2013, http://zythophile .wordpress.com/tag/chevallier-barley/.

6. Brian Forster, "Mutation Genetics of Salt Tolerance in Barley: An Assessment of Golden Promise and Other Semi-dwarf Mutants", *Euphytica,* 08-2001, Volume 120, Issue 3, (Dordrecht, Netherlands: Kluwer, 2001) 317-328.

10

Malt Quality and Analysis

"I have learned far more about malt analysis when dealing with screwed-up malt than when things are chugging along."

Andy Farrell, Bell's Brewery

Knowing that quality grain makes quality beer, experienced brewers often encourage newer brewers to spend time on malt analysis, even if the minutiae of malt characteristics seem unimportant in the grand scheme of brewing. This chapter explores the complementary subjects of malt analysis and specification, and the practical benefits of carefully analyzing malt.

Malt Analysis

The wide variety of brewing techniques and styles means there is also a wide range of opinion on what elements of malt analysis are important to (or required by) the brewer. Although important information can be revealed from a close reading of a Certificate of Analysis (COA), not all of the data on this form is useful to every brewer. A homebrewer might be concerned only with color, while a large industrial brewer likely focuses on potential extract. Practically, a COA serves two purposes: to document production in the malthouse, and to predict the ultimate performance in the brewhouse.

The Malt Company - Townville, WI 55555
(555) 555-1234

Certificate of Analysis

Customer	Shipping Date	Tracking Number	Product
Your Brewery	2014-03-11	TR9370	2-Row Brewer's Malt

Bushel Weight	Shipment Weight		Lot Number
42.75	21,068		109072

Crop Year	Variety	Percent
14	Copeland	40%
14	Metcalfe	40%
13	Harrington	20%

Fine Grind Extract, As Is:	78.8	Diastatic Power:	156
Find Grind Extract, Dry Basis:	82.0	Alpha Amylase:	64.2
Fine/Coarse Difference:	1.3	Total Protein:	11.1
Coarse Grind Extract, As Is:	77.50	Soluble Protein:	5.33
Moisture:	3.90	S/T Ratio:	48.0
Color:	1.60	Beta Glucan:	58
		Viscosity:	1.43
Assortment, 7/64:	65.10	FAN:	219
Assortment, 6/64:	26.70	Turbidity:	6
Assortment, 5/64:	6.20		
Assortment, Thru:	2.00	Mealy:	100.00
		Half:	0.00
Friability:	85.5	Glassy:	0.00

Fig. 10.1 Base 2-Row Brewer's Malt mock Certificate of Analysis (COA)

The malt needs of brewers vary based on experience, size, and goals. Some homebrewers might be satisfied simply learning about and understanding the basics. Other brewers need to be able to determine exactly how much malt is required to achieve a desired color or flavor. A professional brewer producing less than 10,000 bbls may judge malt quality by lauter performance and overall consistency. A craft focused regional brewery likely focuses on malt attributes that result in repeatable fermentation performance for consistency in the final beer. Large industrial breweries have their own focus and requirements; the quantity of enzymes needed to convert adjunct brewing materials, FAN, and extract potential topping their lists of malt needs. Flavor is important for everyone across the brewing board, but how the subjective idea of "flavor" gets defined depends on the individual brewer, brewery, and beer.

Most senior-level brewers request and review the COA for each shipment of malt before it is used in the brewing process. It is important for them to understand which attributes vary with the malt type; for example, caramel malts have no diastatic power, and therefore do not need to be tested for it.

The amount of information in a COA can seem overwhelming at first. Much of the data is not clearly defined, and there is no obvious explanation of what the data really means for beer production and quality control. In some cases, the interrelationship of various values is more important than the individual tests. Because individual barley varieties perform differently during malting, values that would appear to be excessive for one variety might be well within the accepted range of another. Although a single COA will reveal something about the character of a specific malt, subtle production variations that occur over time can only be teased out through comparison of different batches. The test results listed on the COA further complicate the document. Although individual malt testing laboratories can be highly precise, accuracy between labs is not always assured, meaning two different COAs for the same malt type might not perfectly match.[*]

Some elements of a COA (like kernel assortment) deal with physical characteristics of the malt sample. Other attributes (like color) are derived from measurements taken from wort made from the malt in a standardized "Congress Mash." This mashing standard, originally developed in 1907, uses very finely ground malt to make a thin liquid which undergoes an intensive mashing routine. After the mashing, the wort is tested for brewing-related attributes such as total extract, conversion time, pH, and runoff time.[1]

Because the Congress Mash happens under very different conditions than the malt will actually be used, some brewers feel that better tests should be developed to predict actual brewhouse performance. To counter the

---

[*] Splitting a sample and having it compared between two different labs can demonstrate this directly.

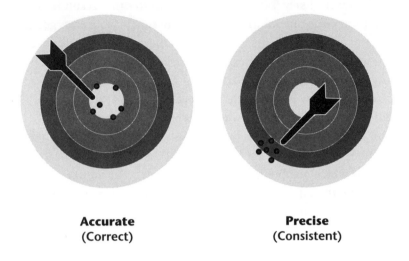

Accurate
(Correct)

Precise
(Consistent)

Fig. 10.2: Although sometimes used synonymously, there is a difference between accuracy and precision. Accuracy is the nearness of a measurement to the true value. Precision describes how reproducible and repeatable several measurements are to each other.

drawbacks of a more than 100 year old mash test, some new technologies provide more reliable data for attributes like enzyme levels and kinetics.[2] Despite its shortcomings, the Congress Mash is well established and understood, and still used to measure malt attributes across the globe.

Throughout a career spanning 43 years, (including a few years as Director of Brewing Raw Materials for Anheuser-Busch) Joe Hertrich has reviewed many, many malt reports and argues that the analytical attributes described in the COA can be consolidated into five basic categories.[3] These "bundles," (as Hertrich calls them) are: carbohydrate modification, protein modification, carbohydrate enzymes, carbohydrate extract, and color/flavor.

Carbohydrate Modification

Carbohydrate Modification describes to what extent the carbohydrate fractions of the barley have been broken down. This category includes beta glucan level, friability, viscosity, and fine/coarse difference. Beta glucans are gums comprised of long chains of sugars that are not broken down by mashing enzymes. High beta glucan levels cause filtration problems, both in the lauter tun and the finished beer. Elevated levels of gums also increase the viscosity of wort. Friability describes how easy a substance crumbles when pressure is applied (highly modified malt is very friable). The fine/coarse difference is determined by comparing extract yields when worts are made from malt that is both finely ground (virtually powdered) and coarsely ground (as in a standard mashing procedure). Because poorly modified malt doesn't release extract easily, it will have higher fine/coarse difference than well modified malts.

Protein Modification

Protein Modification covers the soluble/total protein ratio (S/T), Free Amino Nitrogen (FAN), and pH. As barley is modified into malt, proteins in the grain are broken down into smaller fractions. These fractions are easily solubilized into wort, and the amount of overall malt modification can be measured by the ratio between soluble and total proteins. This ratio (also known as the Kolbach Index) changes greatly between barley varieties. The S/T values themselves can also vary widely. A Belgian brewer making pale strong beer may want malt as low as 38 S/T, while a brewer of a light American adjunct lager may desire S/T values of 46 or greater. The malt used for the latter is high desert barley; very plump and bright with a low protein level. It needs to be highly modified to get sufficient enzyme development for that specific style of beer.

In essence, the higher the S/T, the greater the level of modification of the malt. There are differences in S/T values between barley varieties for fully modified malt, so S/T doesn't always tell the whole story.

FAN is a measure of amino acids, the smallest protein fractions. Although sufficient FAN is required for healthy yeast growth and fermentation, excessive quantities have negative effects on beer quality. Because brewing adjuncts do not contain FAN, worts with high adjunct-to-malt ratios need to get all the FAN from the malt fraction of the mash. Conversely, all malt worts generally provide adequate FAN levels for brewing purposes. Grain pH drops during malting and can also be used as a measure of modification. According to Joe Hertrich's "*Malting from a Brewer's Perspective*" talk given at the 2007 Craft Brewers Conference, pale malts should have pH values between 5.92 and 5.99; levels above or below this can indicate under or over modification.

A Tale of Two Malts

European malt destined for all malt beer has a very different set of analysis data than North American malt optimized for making adjunct beer.

European malt may be made from a barley variety that has low total protein (8.5 percent) and is fully modified at 38 percent S/T. North American malt values may be 13 percent and 50 percent respectively.

If wort of the same strength were made from both, the European malt would release 3.23 percent protein (38 percent of 8.5 percent) whereas the value for North American malt would be 6.5 percent. That would seem to be a big difference, but when the North American malt is used to make beer with 50 percent adjunct (which gives no FAN) the effective rate drops to 3.25 percent (50 percent of 6.5 percent).

In the end, the FAN values for these two different worts are virtually identical. Both malts are well suited to make these respective worts.

Barley variety is what drives both total protein and S/T for fully modified malt and in my opinion, its importance cannot be overstressed.

Carbohydrate Enzymes

Carbohydrate Enzymes reports on a COA measure two things: alpha amylase (AA) and diastatic power (DP). AA measures the quantity of the alpha amylase enzyme only, while DP (expressed in degrees Lintner or °L) includes both alpha and beta amylase content. Although higher levels of these enzymes indicate a greater degree of modification, decreased levels can be indicative of denaturing from higher temperature kilning. Some six-row base malts measure at ~180°L, but only ~40°L (the level seen in some standard English pale ale malts)[4] is required to efficiently convert a mash to work. Malts with high DP are sometimes referred to as "hotter" and can be difficult to control when used in an all malt mash. In Europe, DP is expressed in "Windisch-Kolbach" units, where 1 WK=(°L * 3.5)-16.

Carbohydrate Extract

Fine Grind Dry-Basis (FGDB), Coarse Grind As-Is (CGAI), moisture, total protein, and assortment all relate to the Carbohydrate Extract category. If separated into the components of soluble carbohydrates (extract), insoluble substances (such as husk and spent grain), water, and protein, the percentage of the soluble carbohydrates (which are responsible for the flavor and alcohol in beer) will decrease if any of the other categories increase. Because CGAI includes malt moisture, it represents predicted extract yield for malt when used in a normal brewhouse situation. A Dry Basis value has the moisture factored out. This is used by some brewers to compare potential extract between multiple lots. Proper moisture levels are important because low moisture malt is more susceptible to breakage in handling, and high moisture (or "slack") malt can spoil in storage. The moisture percentage also carries financial implications, as the weight of the moisture adds to the final weight that dictates malt prices.

Color and Flavor

Maillard reactions drive color and flavor development simultaneously, so in some cases, but not all, darker color indicates more "malty" flavor. Due to the subjective nature of taste, flavor as reported on a COA is not particularly useful; it would be surprising if the flavors listed were anything but "malty." Malt color is reported in Standard Reference Method (SRM) or European Brewery Convention (EBC) units. SRM is based on the Lovibond scale and determined from a Congress Mash wort analysis. This spectrophotometric measurement does not indicate hue, only depth of color when viewed at a wavelength of 475 nanometers.

Additional Items of Importance

In addition to the analytical values, the COA sometimes reports on other notable attributes. These additional factors aren't listed on every COA, but are important to the malt savvy brewer nonetheless.

Assortment

Assortment indicates the homogeneity of the barley used to make malt. Because malting and milling are batch processes, neither can be optimized when the size of kernels is highly variable. A size measurement is performed using a shaking device equipped with successively tighter screens (screen sizes on a COA are typically listed in 64ths of an inch in the US or millimeters in the UK). Large, plump grains remain on the top screen and thinner, smaller kernels drop through to tighter screens. Plump kernels generally have a higher ratio of endosperm and produce better results for the brewer.

Bushel Weight[*]

Bushel weight is a bulk density measurement: weight per unit volume (a bushel equals 1.244 cubic feet or about 9.3 US gallons). A higher bushel weight is usually better.

Hartong Number

The Hartong Number measurement is mainly used in Europe. This test compares extract percentages derived from mashing at five different temperatures. It is seen as an indicator of malt modification, yield, enzymatic power, and amino acid availability. Typical values would be 35 to 39.

Deoxynivalenol (DON)

Deoxynivalenol is a mycotoxin produced by certain grain molds (for more information, see Chapter 6). It can cause gushing problems in finished beer and is regulated by the FDA. It is also known by the rather self-descriptive synonym "vomitoxin" for its detrimental effects on animal digestive tracts.

Nitrosamines *(NDMA)*

Nitrosamines are carcinogenic and result from the kilning of malt using direct exposure to combustion gases (as noted in Chapter 2). These should not be present in high quality finished malt.

Acrospire Length

Growth or Acrospire Length is occasionally reported on a COA, and describes how far the acrospire has grown. It is one of the oldest measures of malt modification and easily assessed visually. In general, modification matches acrospire length, and fully modified malt will have an acrospire that is at least ¾ the full length of the kernel.

Mealy/Half/Glassy

Mealy/half/glassy is mainly used in conjunction with COAs for caramel malt. It is assessed by cutting the malt kernel to examine the internal structure. Glassy malts are created when starch converts to sugar and is then dried. As such, base malts should be mealy, while caramel should be glassy. Malt that has both glassy and mealy components is reported as "half."

---

[*] This is not the same as the bushel unit of weight which is exactly 48 lbs for barley, 34 for malt.

Broken Kernels

Broken kernels and foreign seeds are indicative of poor grain handling practices. Quality malsters will do their best to mitigate both of these undesirable traits.

All too often, information about barley variety and crop year do not make it onto the COA. Many brewers feel that this is critical information that needs to be included, and are working to make future COAs more comprehensive. What is reported on a COA is mutually determined by brewer and maltster, and many brewers don't know to ask for this information.

What Brewers Need to Know About Malt

"Rather than add more analyses to the list, perhaps we should focus on fewer, looking for information, rather than data. Ultimately, it seems, a brewer's judgment on malt quality, using malt analyses as an aid rather than a straightjacket, is likely to be the best path."[5]

–Bill Simpson

Deciphering the information in a malt COA takes practice. Much of the insight into how malt will perform in the brewhouse relies on the complex relationships and balances between analytical values. Determining which malt performs well and produces the best beer requires a deeper level of communication with the maltster that goes beyond the interpretation of a paper report.

One of the most significant gaps in analysis comes from the necessity for averages. The aphorism about the man with his feet in an ice bucket and his head in an oven captures the sentiment well; "on average" he was at a comfortable room temperature. Most maltsters manage process deviations by blending different malt pieces together to reach a desired average. Even though the data on the COA appears correct, this manipulation can create malts that while statistically similar, will perform quite differently in the brewhouse.

When asked about COAs, Dan Carey of New Glarus Brewing Company replied that he doesn't trust them in general, because errors in some test methods mean the document won't meet the precision requirements demanded by some brewers. But, he cautioned, do not rely too much on the beta glucan test. In conversation with a well-respected maltster, Carey learned that from the maltster's standpoint, the test isn't accurate enough and doesn't tell the brewer the size of the beta glucan molecules (an important thing to know when formulating a recipe). He noted that the COA, when used as a control parameter for particular varieties, can be a helpful tool for discussion even though it may not be accurate between laboratories. "When I see a Metcalfe BG of 70 ppm, I know I will have a nice lauter and extract. When I see 130 ppm, I know extract will drop a bit and I might have a second deep cut."

Carey hasn't found a great consensus between maltsters. For example, one malthouse's 2.2°L may be another maltster's 1.8°L. "If you want consistent malt, buy from a reputable maltster and specify variety. Don't get hung up on COAs," Carey advises. "If you ask me about the quality of a given shipment of malt, I'll tell you to brew three batches and you'll know."

Maltsters view malt analysis and specifications through a very different lens than brewers. To them, the S/T ratio and beta glucan levels drive malt quality, and many maltsters use these two variables to gauge steeping and germination process control. They will also pay particular attention to friability and diastatic power tests because sufficient protein degradation and enzyme development means they have produced very desirable brewing malt.

For Dave Kuske of Briess, the most important information on the COA for him is (in descending order): assortment, color, extract coarse grind, bushel weight, friability, glassy/mealy, viscosity, diastatic power, AA, moisture, turbidity, BG, S/T, FAN, and FC differential.[6]

Certificates of Analysis—Important Information for Brewers

While acknowledging the limitations, it is important to recognize that the COA *does* contain information valuable to the brewer, especially when comparing malt lots made from the same variety, the same crop year, and the same supplier. In the examples that follow, it should be assumed that the terms "higher" or "lower" are comparative to same variety/crop year/supplier.

Inadequate Protein Modification: Malt with inadequate protein modification will show lower S/T values than normal. Performance in the brewhouse will suffer with slow, difficult lauters, and poor extract recovery. The wort, although turbid, may be bland. The resultant beer may suffer from erratic attenuation, poor physical stability, and difficulty in filtration.

Low Carbohydrate Modification: Elevated fine/coarse difference may hint at low carbohydrate modification. Because carbohydrate and protein modification are closely linked, it is not surprising that low carbohydrate modification and low protein modification often present with similar symptoms. Both produce steely ends (the rock hard, undermodified distal portion of the grain), which are abundant sources of native beta glucans. These dense sections are problematic during milling, and the carbohydrates encased within them can be difficult to extract.

Protein Overmodification: Excessive consumption (and thus loss of potential brewing extract) of carbohydrates during the malting process leads to overmodification. Higher S/T values may indicate elevated protein modification, which results in excessive breakdown of foam positive protein fractions, and negatively affects the beer's body. Both the cell structure and the small starch granules will be fully broken down in overmodified malt.

For most brewers, low S/T levels are more problematic than high ones. Although not generally an issue for all malt brewers, low FAN levels in wort present problems like insufficient yeast nutrient, difficult fermentations, and increased sulfur production for adjunct brewers.

Conclusions

It is important to recognize that there will be significant variability in malt across barley varieties, and between crop years for the same variety. Individual grain lots can behave very differently in the malthouse and mash tun. These changes can cause significant quality concerns in both the malting and brewing processes, but they can be mitigated through an open dialogue between the brewer and maltster.

Farmers, maltsters, and brewers depend on agricultural raw materials that are heavily influenced by annual variations in the weather. A less than ideal situation can only be overcome with a thorough understanding of the critical parameters for every player in the game, and the needs of each party can be addressed more effectively by working together.

It is important for a brewer to develop a solid relationship with her maltster, and speak with them regularly. Consistent problems should be pointed out so the malsters have a chance to rectify any quality issues. It is also good practice to document what malt was used to achieve the best batches of beer, and seek consistency and repeatability when choosing malt for future batches.

Open communication leads to trust, which allows a maltster to best use their full range of skills to malt specific varieties. A critical component of that dialogue is the ability (of both parties) to communicate and accept information about developing problems. Common sense would dictate that if the brewer is going to have major concerns about every little variation, the maltster has little impetus to mention any problems. Brewers should strive to let the maltsters do their jobs, but need to remain alert and accountable for the quality of grain that goes into a mash.

As part of any contract, specifications for the malt are necessary. The art lies in defining critical components without putting excessive demands on the maltster. For example, even in a drought year that produces barley with elevated protein contents, a maltster should be able to meet a specified maximum soluble protein level. However, those protein levels would likely mean elevated beta glucan levels, too, which may go against the brewer's specifications. Because color is also affected by protein levels (higher promotes darker), it too could become unintended collateral damage in this scenario, leading to a malt that meets one of the brewer's criteria, but completely misses others.

There are a number of published works devoted to finding and appropriate level of specification between brewer and maltster. At a high level, there seems to be a consensus about the need for perspective and dialogue. In a 2001 paper presented in Canberra, Australia, Dr. Bill Simpson of Cara Technology noted that:

"Variability is at the heart of the problem. Variable malts themselves are not particularly difficult to handle, except in extreme circumstances. What causes more problems is blending of different malts by maltsters or brewers to give a single batch that is 'in specification'. While superficially the practice makes sense, it can cause more problems than it solves. What is needed is more understanding between maltsters and brewers, and more

trust. It is a brave maltster who dares to convince a brewer that they should accept an out of specification malt, on the basis that it will outperform an in-specification, but nevertheless 'rogue', batch. Some companies have moved toward this co-operative ideal, but there is some way to go."[7]

Author's note: I would be negligent in my duties as a brewer if the parting advice for this chapter about malt analysis, specifications, and brewing performance did not include the often repeated admonishment to always "blame the maltster." Brewers with limited understanding of malt have used this maxim enough times that it must be true by now. And if you believe that, I have some magic hops I would like to trade you...

References

1. Roland Pahl, "Important Raw Materials Quality Parameters and Their Influence on Beer Production", Presentation at the Bangkok Brewing Conference, (Bangkok, Thailand, 2011).

2. Paul Schwarz & Paul Sadosky, "New (Research) Methods for Barley Malt Quality Analysis", Presentation at Barley Improvement Conference, (San Diego CA, 2011).

3. Hertrich, Joe, "Unraveling the Malt Puzzle", Presentation at Winter Conference of MBG/MBAA – District Michigan, 2012.

4. T. O'Rourke, "Malt specifications & brewing performance", *The Brewer International*, Volume 2, Issue 10, (London : Institute & Guild of Brewing, 2001).

5. W. J. Simpson, "Good Malt – Good Beer?" Proceedings of the 10th Australian Barley Technical Symposium, (Canberra, Australia, 2001).

6. Dan Bies & Betsy Roberts. "Understanding a Malt Analysis", *The New Brewer*, (Boulder, CO: Brewers Association, Nov-Dec 2012).

7. W. J. Simpson, "Good Malt – Good Beer?" Proceedings of the 10th Australian Barley Technical Symposium, (Canberra, Australia, 2001).

11

Malt Handling and Preparation

The main economic forces at play in the northern reaches of Michigan's Lower Peninsula have traditionally been lumber and recreation. Bellaire is home to both the spectacular Torch Lake and in more recent years, Short's Brewing Company. The relaxed and quirky local vibe at the pub is reflected in the seating choices; couches constructed out of salvaged Cadillac seats mounted on recut oak stock, and bar benches made from repurposed school bus seats.

In 2005, Joe Short received a call from an equipment vendor who was dismantling a brewery downstate. Was he interested in some free malt? As a young brewer who had made a whopping 178 barrels of production the prior year, Short's Brewing was "strong of back, weak of bank" and Joe jumped at the chance.

Using the brewery's entire staff as labor, the whole-kernel malt was removed, a bag at a time, from the shuttered bulk system and transferred into used malt bags. Joe Short recalls that the diligence and neat work performed by one of the pub tenders caught his eye. "I loved the way that Leah closed and stacked those bags." Her tender care of the gifted malt moved Joe so much that eventually their professional relationship turned into a romantic one, which eventually blossomed into a marriage. Who knew that unloading malt could stir a romance in the heart of a brewery? Short's malt gift story is only a small example. Regardless of the size of the brewery, some form of malt handling occurs before the beer is ever brewed. Transport, storage, and preparation are less compelling but still very important aspects of the brewing world that no brewer should ignore.

Packaging

Boston's Commonwealth Brewery used English malts in the late 1980s. At that time, one of the tasks for an entry level brewery worker was to unload containers of malt. Much of the grain was packaged into 50kg (110 lb) bags and floor-stacked in a shipping container. Receiving a single shipment required sorting and shifting 50,000 pounds of malt into storage at the brewery farm in less than two hours, to avoid any waiting charges from the trucking company. More than one brewer found that this intensely physical task was a sufficiently powerful disincentive to drinking too much the night before. In those days, noticeably poor performance led to ridicule and brewery-wide chatter. Building character notwithstanding, bulk handling of malt and spent grain was a seriously labor-intensive task that brewers might have poked fun at, but desperately needed.

The production level of the brewery often dictates the method of packaging and transport. Homebrew supply shops offer both prepackaged bags and "scoop your own," while large brewers order by the international shipping container load. Although malt is fairly stable, the level of packaging has to be sufficient to protect it from insects, moisture, or off-flavors.

Malt is packaged into different sized bags, either at the malthouse or an offsite warehouse. The bag material needs to protect and contain the malt, and all manner of papers and plastics are used to contain the grain. Hundreds of years ago simple woven burlap was used, but it did not provide much protection from moisture or pests. By and large, the unwieldy 110-pound bags have been replaced by 50-pound or 25kg bags today. Regardless of size, proper lifting technique is necessary to prevent worker injury, and to prevent a condition called "Brewer's Back." Looking back at the Commonwealth Brewing cohort, those old-school, side-bunged Hoff-Stevens kegs that seemed to have been purposefully designed not to be picked up by normal humans could have contributed to all the chiropractic problems of early brewery workers.

Pallet sized bags called "super sacks" can hold up to a metric ton (1000kg or 2205lbs) of malt. They can be filled with a single type of malt or contain a blend for ease of use at the brewery. Given the massive size and weight, these bags must be moved by forklift, chain hoist, or other more specialized equipment. If less than a full bag is to be used for a brew, some form of metering or measurement is also needed.

For many craft brewers, bulk malt will be conveyed into an on site bulk silo. The most common method is to use a pneumatic blower truck. Improper handling at this stage can result in significant damage to the malt, so careful unloading is integral to retaining quality and minimizing loss.

Intermodal shipping containers (developed in the 1950s to facilitate efficiency in moving freight through a supply chain utilizing truck, rail, and ship) are sometimes used to move malt transcontinentally. The common size for both the 40' and 20' long containers is 8.5' tall and 8' wide. Generally both a food grade liner and door-side bulkhead are placed inside of the container. Bulk malt is placed directly into the assembly and the container is sealed for shipment. The grain can be unloaded by vacuuming or by tipping the entire trailer to allow the malt to pour out through a doorway in the bulkhead.

Hopper-bottom type bulk trailers can efficiently carry large loads of grain and are generally limited to a 50,000 pound weight restriction in the US. After loading the grain through the open top, a cover is added to protect the load during transport. Sliding gate valves at the bottom of the trailer allow the grain to be emptied at the destination. On a large scale, railcars are the most common method for moving barley and malt; each car can hold about 180,000 pounds. For very large loads traveling long distances, bulk malt may be moved using barges or other large ships.

Receiving

Regardless of how the malt arrives—by postal mail or larger means—it needs to be inspected. Damaged packaging is an obvious and common concern, but there are other shipping related factors that can harm the malt, such as excess moisture leading to spoilage in the form of mold growth. Advances in supply chain management notwithstanding, mistakes in the ordering process, product picking, and shipping problems can result in different malt being delivered than what the brewer ordered. A proactive brewhouse manager should review and confirm the purchase order and shipping documentation against the delivery, as soon as malt arrives.

Bulk deliveries will commonly have a shipping seal that verifies the load has not been tampered with. Removing these seals with the delivery driver present provides an excellent opportunity to assure that unloading is done properly. Mishandling of the malt by trucks equipped with pneumatic equipment can cause significant damage (like separation of the husk material from the grain). Pneumatic systems are used for moving a wide variety of bulk materials such as flour, sugar and even plastic pellets. Malt has significantly different flow and handling factors than powdered products, and an inexperienced delivery driver who typically transports items other than malt may accidentally causes problems for the brewery.

In pneumatic systems, air blows through a hose that connects the brewery silos to a pipe attached to the bottom of the truck hoppers. In the case of flour-like material, a low material flow is introduced into high-volume airflow, to prevent blockage in the transport tubing. This "dilute phase conveying" method results in high product velocities. Friable substances (like malt) can be damaged as individual kernels collide with any protrusions or bends in the pipe. An observant brewer may notice a greater percentage of husk material and chaff as the final quantities of malt are removed from the silo. As it is less dense, this material tends to separate and move to the top of the mixture. Even with perfect lautering practices, husks and chaffs are low in extract, high in tannins, and poorly suited for brewing.

The better alternative for handling friable substances is "dense phase conveying," which introduces larger quantities of material into a slower moving airstream and as such, will minimize the damage to the product. Convincing a delivery driver who is unfamiliar with malt to attempt dense phase conveying can be a challenge for even the most persuasive brewer, as their experience with a product like flour may convince them that low airflow will clog up a transfer piping system.

Malt receipt is an excellent time to check the Certificate of Analysis (COA) for any abnormalities. A conscientious brewer should always ask himself: "Is this the malt I expected (and hopefully ordered)?" Additional information on the importance of and how to interpret a COA is covered in Chapter 10 of this book. It is considered good practice to pull a sample of bulk grain at time of delivery. Grain sampling triers (See Figure 11.1) can be used to pull material from deep in the shipment of grain to get a sample representative of the whole batch.

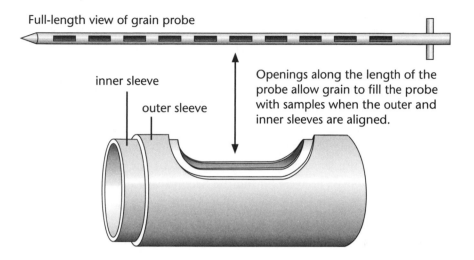

Full-length view of grain probe

inner sleeve

outer sleeve

Openings along the length of the probe allow grain to fill the probe with samples when the outer and inner sleeves are aligned.

Fig 11.1: The use of a grain sampling trier allows sampling that better represents an entire batch.

Bulk Grain Safety

Handling of bulk grain poses some serious safety concerns. Each year, many agricultural workers are killed by collapsing grain beds, making proper training, supervision, and personal protective equipment of paramount importance. Material flow problems are commonly encountered in large-scale grain storage and processing operations. It is important to recognize and respect the hazards that seemingly innocuous grain storage bins represent. Cavities can form deep in a pile, and when they collapse the flowing grain can quickly swallow someone working on top of the pile. Likewise, material that is flowing to an outlet can easily pull a human into the stream. Once engulfed, it is very difficult to escape the downward flow without assistance. Because these flow conditions exist even in hopper-bottom grain delivery trucks, it is necessary to protect workers from these hazards by prohibiting entry to the grain. A sobering fact that highlights the need for industrial awareness and safety: the majority of confined space fatalities are persons who were attempting to rescue the first person who wasn't wearing proper safety equipment or following correct procedures.

Storage

Years ago, when inspecting brewing equipment for sale at a shuttered Swiss brewery, I had the opportunity to check out the malt storage equipment. This part of the brewery seemed oddly oversized in comparison to the brewhouse. The many, many silos were arranged in a matrix of walls and conveyors. When questioned about the scale of the operation, the host commented that it was typical of Swiss breweries; a law that required the brewery hold a one year supply of malt was active when the brewery was built. The law was enacted in direct response to the supply disruptions experienced by Swiss brewers during the Second World War. Based on the Swiss example, holding malt for long periods of time can be done effectively, but comes with its own set of problems.

Typically, malt arrives at the brewery with about a four percent moisture level, and it is a fairly stable material. At this moisture level, enzymatic activity is dormant and the soaking, germinating, and kilning processes have minimized the possibility of microbial and insect-based spoilage. However, malt can still experience quality problems after arrival. Assuring that the good malt that arrived at the brewery remains good is the duty of the brewery. Because ambient storage conditions vary widely in both moisture and temperature, what is appropriate for protection in one area of the globe may not be sufficient in another. Higher moisture or temperatures are more problematic for the malt.

As a grain, malt is an intensely concentrated food source for much of the animal kingdom. Birds, rodents, and insects are all attracted to malt for its nutrient value, and if allowed to gain access can quickly infest an area. Sanitary design coupled with regular cleaning minimizes possible problems. Bagged materials should be stored off the floor and away from walls to keep pests at bay. Although acrobatic rodents may still be able to gain access, maintaining a clear space allows for monitoring animal activity and gives plenty of room for the installation of traps and other control measures if needed. Properly sealed grain bags (as they come from the manufacturer) will generally deny insects a chance to get to the grain. Rodents can easily penetrate such packaging and contaminate the malt with urine and feces. It goes without saying that few beers would be improved by these additions. Compromised ingredients must be destroyed as soon as discovered.

Due to their mechanical complexity, bulk operations, milling, and conveying equipment require constant vigilance, upkeep, cleaning, and monitoring to assure that spilled or retained malt does not attract pests. This is especially the case in warm and moist climates. The conditions in a silo located north of the 45th parallel during winter are not nearly as hospitable to bugs as one in a steamy, equatorial swamp. Guarding against infestation by pests such as Indian Meal moths, Confused Flour beetles, and cockroaches begins with comprehensive cleaning programs; if potential food sources are minimized, then insect activity will be as well. It is always easier to prevent an infestation than to remove one, and nothing attracts pests like available food.[*] Good brewery practice also dictates that silos should be emptied and cleaned regularly. If bins are never fully emptied, pests (and their offspring) can carryover from one fill to the next.

An outdoor delivery with spilled grain is an ideal way to attract unwanted bird life, and waste brewery money. Failing to set a high standard to both prevent and clean up spills will signal to the local avian population that "dinner is served" at the brewery. Birds indiscriminately soil the buildings and grounds, which negatively affects both the aesthetics of the brewery and quality of the beer.

Moisture and malt do not play nicely before the mashing process. The enzymatic conversion of starches to sugars provides an ideal food source for molds and bacteria, and controlling wet conditions in grain storage is critical to preventing abundant microorganism growth. Rotting grain develops a powerfully pungent aroma that readily spoils beer and wort. Moisture is also detrimental to the steel components present in most grain handling equipment. Any evidence of excess moisture in the malt should be fully investigated and resolved at the earliest opportunity. Since it is also possible for malt to pick up off aromas from poor storage or transport conditions, a brewery should do its best to protect the malt from any solvents and strong food aromas.

As fats and oils age, they oxidize and develop rancid flavors. Grains such as oats and corn have higher oil contents than barley and are thus more susceptible to degradation. Because of this, inventory of non-barley grains should be turned over more rapidly. Some brewers also feel that caramel malts have a shorter lifespan and should be stored in smaller quantities than base malts to maintain optimal flavor expression.

---

[*] With the possible exception of sex, that seems to be a substantial attractant too. And I think everyone would agree that adding beer to the mix doesn't dissuade them either.

Conveying

Breweries that handle grains in bulk require some kind of conveying system. Grain can be moved through the brewery using an assortment of equipment, including screw augers, bucket elevators, chain discs, cable conveyers, and paddle troughs. Each system has inherent attractions and drawbacks, and not every system is the same for each brewery.

Screw Augers

These large augers move material by turning a screw inside of a pipe or trough. Many screws have a rigid central axle; however a slinky-like helix can also be used. Close tolerance between the screw and side wall is necessary to properly clean the conveyor. There is a limit to the vertical angle (and in turn height) screw augers are able to effectively reach. This angle varies with a number of design factors but staying below 45 degrees seems to be the most agreed upon design.

As with most conveying systems, high speeds tend to increase transport damage, so awareness of rotational speeds is necessary when specifying a design. Flexible augers have the unique ability to accommodate both straight and curved sections and are typically used by smaller breweries with tighter space constrictions.

Bucket Elevators

Bucket elevators are an excellent solution for gently moving malt. "Buckets" affixed to a vertically oriented, continuous belt, scoop material from a bottom trough. At the top of the installation, as the belt turns to start its descent, the material is thrown outward by centripetal force and funneled away by a discharge chute. Although these systems are sometimes used in very large industrial applications and are thus capable of extremely high flow rates, more compact systems are also available for smaller breweries.

Disc Conveyers

Both chain disc and cable disc systems work by pulling a series of molded plastic pucks through a continuous piping system. Material introduced into the loop is pushed forward by these discs. This method allows for gentle conveying both vertically and horizontally. At the installed corners used to change direction, there is generally an enclosed rotating

Fig. 11.2: Well laid out grains area at Starr Hill Brewery in Crozet, Virginia with batching bin, 4-roll mill, chain-disc conveyor, and fabric malt storage silos. A Feed-Pro batch controller is mounted above the mill.

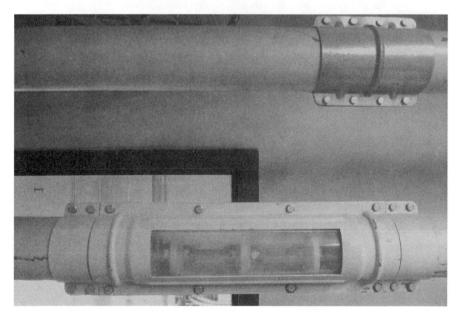

Fig. 11.3: Closeup of chain-disc conveyor.

wheel to minimize friction and equipment wear. Relatively inexpensive, agricultural-grade, small-scale solutions are available, as are larger versions of European manufacture.

Plastic paddles pulled by a continuous chain loop through a formed metal trough can also be used to move grain gently. Slide gates mounted flush to the trough bottom allow for multiple, cleanable discharge points. Generally the entire chain loop is enclosed in a single trough with the paddles returning to the main grain source above the moving grain.

Pneumatic Systems

Permanently installed pneumatic systems are also used for grain movement in some breweries. Essential components of these systems include airlocks (used to separate the conveying loop from static bins or equipment), cyclones (needed to separate conveyed material from the moving airstream), and some type of blower (to provide force for the overall system). Proper engineering is necessary to achieve needed mass flows with minimal product damage. Gently sweeping, long-radius piping bends are also required. As

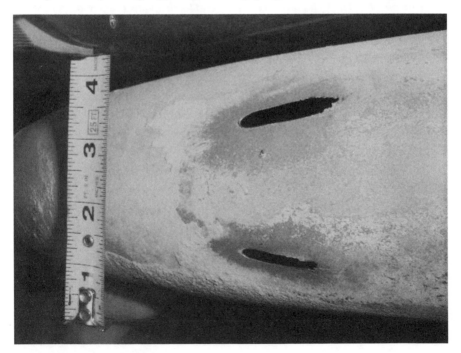

Fig. 11.3: Damage caused by malt to a pneumatic conveying pipe.

with truck-based pneumatic systems, both dilute and dense phase conveying is possible. Some systems work via vacuum, pulling the grain forward, while others operate via pressure, pushing it from behind.

Areas that are not smooth in pneumatic conveying systems can be very problematic. This is especially true about the routing used to fill a silo from a bulk truck. Years ago at Bell's, a cam-lock fitting was attached to the silo filling pipe by using screws that protruded inward. The malt was heavily damaged when the rapidly moving grains collided with the screw threads during filling. If the driver or operator does not have experience unloading malt, they might use hard 45 degree or 90 degree angle joints as part of the hose run which can cause inadvertent damage to the malt. Although malt is friable, it is also abrasive, and can erode conveying components over time.

Cleaning

As an agricultural commodity processed in bulk through industrial scale operations, there are ample opportunities for extra, unwanted items to get mixed into the malt. Stories from around the brewing community have described all kinds of alien additions: some brewers have found stones from the field, tramp iron from aging storage bins, and even once, a walkie-talkie, mangled from a trip through auger systems after presumably being lost by a malthouse worker. Cleaning equipment (installed at the malting facility) removes a significant portion of foreign material, but some brewers take additional steps to ensure their grain is free of anything that shouldn't be mashed. Grain mills are often equipped with magnets installed at the malt delivery inlet to remove ferrous particles such as the odd nut, washer, or bolt. These magnets should be inspected and cleaned frequently to assure proper operation; the cleaning frequency that is needed will vary based on malt quality and volume processed.

More advanced cleaning options that separate the malt stream based on size and density are also available. In these systems, particles larger than and smaller than malt are separated by passing the grain across screens constructed of defined mesh sizing. Dust, chaff, and other light density particles are removed by passing malt through a moving airstream. The oscillating gravity table removes small stones and other dense materials. These operations can be done separately or incorporated into a single machine unit termed a combi-cleaner.

Weighing

Regardless of brewery size, some control of malt mass is needed to assure brewing consistency. Unless the formulation is consistently written in whole bag increments, the malt will need to be weighed on scales. For smaller operations, a bucket weighed using a kitchen scale generally suffices. With the massive quantities in a bulk system, some type of in-line weight control is needed.

The simplest system uses a weighing chamber that discharges when a set weight is reached. Batch size is controlled by automatically counting the number of cycles the "dump weigher" goes through. "Load cells" are devices that electronically translate physically applied strain forces (such as weight) into fairly precise mass data. Installing them on the load bearing connections allows a container to become a weighing system and as such they are used in some breweries to weigh bins or batch sized loads. Load cells are also used in the "Feed Pro" weighing system. This equipment uses a horizontal auger turning at a fixed speed to measure feed rate. As an example, if grain takes 10 seconds to move through the weighing section and the weight of the grain in the section is 20 pounds then the flow rate equals two pounds per second. As a result of the simple mathematical connection, flow rate and total weight are able to be easily controlled through the system.

Dust Control

As the friable malt is conveyed or ground, some dust is created. There are few substances in the brewery that carry a heavier percentage of wort and beer spoilage microorganisms than malt dust. Carryover into either fermentation or packaging operations can cause significant cleaning, sanitation, and microbiological problems.

As noted earlier, insects and other vermin are attracted to the ample food source that grains provide. Prevention of dust buildup reduces the likelihood of infestation. In addition to the brewing and contamination issues, excessive dust is unsightly and can contribute to the perception that the facility is poorly run.

Areas that have noticeable airborne dust are unpleasant and unhealthy to work in. Obvious respiratory factors aside, the most notable hazard from excess dust is, maybe surprisingly, grain dust explosions. Any material that will burn in air in a solid form can be explosive when in a finely divided (powdered) form. An explosion can occur when a sufficient quantity of an airborne combustible powder dispersed within a confined area meets a source of ignition. Although fire, spark, or flame can be minimized, ultimate protection is achieved by controlling the amount of dust in the air.

Dust control can be accomplished in many ways. The first and most obvious is by sweeping, vacuuming, and other manual cleaning techniques. Identifying and sealing areas where grain dust originates helps to minimize the overall dust load. Adequate ventilation can remove airborne dust before it has a chance to settle and present a bigger cleaning problem. Larger breweries are generally equipped with some type of centralized dust aspiration equipment that removes small particles during conveying. The filter separator, or bag house, consolidates the removed material so it can be added to spent grains after mashing, and used for livestock feed.

Grain Dust Explosion Hazards

Grain dust can explode.[1] For a fire to occur three elements are needed: a fuel source, oxygen, and heat. A fire changes from a relatively simple burn to a frightful explosion based on the type of and availability of fuel. Kindling wood starts more easily and burns more quickly than the log it was split from, because the large surface area allows for rapid combustion. Imagine a microscopic fuel source made airborne and captured in a small place. The combustion can become so rapid and intense that it results in a massive explosion capable of destroying a building.

Combustible dust is defined in the OSHA regulations as being below 420 microns in size and "able to burn in air."[2] So how much dust is required to cause an actual explosion? The Lower Explosive Limit for grain dust has been reported as 55 grams per cubic meter. This quantity would be sufficient to significantly reduce the ability to see very far. It is not likely (unless hit with an open bag of flour) that a brewer has experienced these levels of dust concentration. So, where is the hazard for breweries and malthouses? The answer lies in secondary explosions. If a small explosion or deflagration were to occur in a limited area, the resultant shock wave would force any dormant dust into the air. The usual culprit of large scale explosive events is rippling waves of secondary and tertiary explosions.

There are many potential ignition sources that could set off an event. Smoking, welding, or other open flame around grains systems is a very bad idea. Other sources of ignition include sparks arising from arcing electricity, metal striking metal, or electrostatic discharge. The high heat from the friction generated by a misaligned drive belt or an under-lubricated bearing could also create enough heat to spark a fire and subsequent explosion.

Grain Bins

Bulk operations use silos or grain bins to hold received grain until it is needed. Although most are constructed of painted or galvanized mild steel, other materials like plastic are also used. Many are fabricated at a dedicated shop and shipped to the brewery in one piece. Obviously these bins are limited to a size able to be transported over the road, so for larger installations, bolt-together bins may be a better option. Cylindrical containers are an inherently stronger design, and more widely used than containers of other shapes. For high density storage, square or rectangular bins with shared demising walls may be preferred by certain breweries.

Regardless of the physical arrangement, grain will need to be easily removed from the vessel. How the grain flows as it is removed is governed by vessel geometry. Whole kernel malted grains tend to flow easily in bulk systems. Malt has an angle of repose of about 26 degrees above horizontal, so provided the surfaces are reasonably smooth and steeper than 30 degrees, hang-up tends not to be a problem. In practice, most brewery bins are built with bottoms at least 45 degrees above horizontal. When specifying square bins it is necessary to consider the reduced angle where two sides intersect to assure proper emptying.

Milling changes the density and flow characteristics of the malt. Because frictional forces have a greater effect on smaller particles, it is not uncommon to find a rubber mallet* positioned nearby to coax any stubborn grist into the mashing vessel. Options beyond manual beating include pneumatic or electric powered devices that shake the grain bin.

System Cleaning

Grain handling systems should be designed to allow thorough cleaning as material buildup in nooks and crannies of the machines can be problematic. Grain conveying systems are often difficult to inspect because they

---

* No relation to the author, my name has two t's.

are sealed to prevent dust from escaping. In addition to the sanitary issues, problems such as unintentional intermingling of different types of malt can occur from poor system design. Validation of procedures and documentation that assure regular cleanout of the entire system is a necessary prerequisite for organic certification. A rigorous and codified review from this perspective can help to assure that good housekeeping practices are applied and effective in keeping the grain consistent and clean.

References

1. Combustible Dust in Industry: Preventing and Mitigating the Effects of Fire and Explosions http://www.osha.gov/dts/shib/shib073105.html Accessed 3-17-2013

2. A Guide to Combustible Dusts http://www.nclabor.com/osha/etta/indguide/ig43.pdf Accessed 7-7-2013

12

Milling

"Malt newly pulverised is inflammable, and apparently electrical. The destruction of Barclay's brewhouse, London, in 1832, was caused by the accident, that a man happened to lift one of the covers upon the box of the Jacob's ladder which conveys the malt to the hopper, and to thrust a lighted candle amongst the fine powder-like malt that was flying about when the ladder was in motion. Undoubtedly, the dry state of the grain, and its electrical condition, arising from friction in breakage between the rolls, had occasioned a state of gradual decomposition, and brought some hydrogen into the box from the vast quantity of malt that was being crushed at once, and hence its inflammability; but the danger of such an accident does not arise when the process of crushing has been ended, and the gases have been allowed to subside."

- William Littell Tizard,
The Theory and Practice of Brewing Illustrated

Before malt can ever grace the waters of a lauter tun, the starchy endosperm trapped beneath the husk needs to be exposed. This process (called "milling") involves any mechanical action that breaks or cracks the hard husk of dried malt to prepare it for brewing. It may seem like an obvious step or a brewery afterthought, but without milling to bare the sweet result of malting, the hot waters of the mash would have a difficult time extracting any sugars into the wort.

Milling can be done dry or wet, depending on the equipment and requirements of the brewery. Malt is generally milled in-house because once the husk is cracked, the endosperm inside is exposed to natural moisture, reducing its viable shelf life. This chapter will briefly discuss the various methods of milling, and why a brewery might favor one method over the other, and touch on grist analysis during milling.

Dry Milling

After the grain is batched, it is ready for milling; the first operational step of the brewing process. While there are many different types of equipment used by brewers worldwide, the simplest method for the controlled crushing of malt involves passing the grain between closely spaced rotating cylinders. There are simple, affordable, and effective hand-cranked versions available for homebrewing and nano-brewery sized grain bills, and large, motor-driven milling behemoths for large scale brewing.

Before their windfall of free malt during the courtship of Joe and Leah Short referenced in the prior chapter, Short's brewery used only pre-ground malt. The gift of whole kernels (which should have been a financial boon) stressed the resources of the brewery. Faced with the need to grind kernels on a limited budget, Joe and his crew cobbled together a hopper, attached it to a Schmidling Maltmill, and powered the assembly with a secondhand motor. Milling a batch of grist for the seven-barrel system took 90 minutes using this modified homebrewing equipment. Extract yields were not ideal, but since they "were always putting too much [malt] into the mashtun" it may not have mattered.

Although commercial scale machinery can include options and enhancements that allow for greater control and higher throughput, the objective for milling remains constant: controlled size reduction that enables the conversion and extraction of malt components in a timely manner. Defining these factors depends in large part on brewhouse equipment and brewery product, both of which can vary greatly in practical application. For example, the practical milling criteria for a high gravity stout wort made in a nanobrewery's mashtun is quite different than that of an industrial scale adjunct beer processed through a mash filter.

The ideal grist for a typical brewery would have the outer husk material separated from the grain with minimal damage, all while crushing the endosperm to a small but uniform particle size. These goals are inherently in conflict with one another, but with careful milling, are certainly within reach. The brewer must determine how to balance these opposing factors to optimize the milling process for her individual brewery.

The brewer wants the husk material as intact as possible for multiple reasons. The husk contains elevated levels of polyphenolic compounds. Breaking the husk material into smaller fragments increases the concentration of polyphenols in the wort, which can contribute harsh, tannic flavors. Larger husk material provides significant structure and lauter bed porosity, a quality that is essential for rapid and effective extract separation.

When the grist hydrates in the mash, the native enzyme systems in the malt are activated. They are released into (and travel through) the mash water to act upon their target protein or starch substrates. As endosperm particle sizes decrease, the surface area increases, and as a result, the enzymes act more rapidly and thoroughly. This is why breaking the endosperm into smaller fragments generally results in greater extract yield in the wort. Small increases in extract efficiency can result in significant cost savings for industrial scale brewers, but for breweries operating at lower volumes, savings in lautering time may be a far more important.

Regardless of the tradeoffs in milling, consistency is always the desired goal. Uncrushed malt kernels struggle to release any extract, and may pass through the mashing/lautering steps without adding any appreciable sugar to the wort. Malt that is crushed too finely produces dust-sized fragments that tend to impede the flow through the lauter bed, leading to incomplete extract recovery. As odd as it may sound, it is possible, with poorly engineered or maintained milling equipment, to have excessive quantities of un-milled and over-milled malt simultaneously.

Historically there have been many methods used to grind grains. For centuries, brewers have used millstones, the iconic, picturesque equipment used to process wheat or corn into flour or meal. Although modern brewers would likely be disappointed by the quality that results from processing malt in this way, based on writings from the mid-1700s, apparently more modern metal mills did not offer immediate solutions. Citing "an old work by Edmunds, dated 1769," Matthews and Lott state:

"The best method of grinding malt is between two stones by the horse mill; these bruise and break the substance of the corn, whereas the iron mills only cut it in two: the malt thus broken by the stones redily emits its virtue, which the cut malt cannot do so easily, being more confined within its hull." (Matthews and Lott, 1899)

Because of the variety of milling equipment that has been used throughout brewing history, understanding the physical forces involved in this mechanical operation can help to optimize a brewery's process. A very simple mill, where malt passes between two rotating cylinders, uses compression forces to crush the kernels. The more friable the malt, the easier it is to process through this type of equipment, and mills require uniform, well-modified malt to output quality grist.

Additional grinding forces and equipment configurations can increase the quality of a brewery's milling operation. By rotating the cylinders at different speeds, milling machines can tear and sheer the malt, and driving one roller at a speed 50 percent greater than the other is not an uncommon practice. Incorporating grooves or fluting into the surface of the rollers provides an even greater shearing effect. For efficient and effective milling, the gap between the rollers should not be completely filled with kernels. Malt should be added to the mill in a controlled, metered way to ensure there is enough space for through and clean milling.

Dividing the crushing operations into multiple steps allows for greater control in the quality of the final grist. In many mills, as the malt transfers from one set of rollers to the next, it is passed over vibrating screens where the crushed malt is segregated by size. A malt kernel may first be roughly cracked in the top rollers and then as moved across a sieving screen where the coarse endosperm detaches from the minimally damaged husk. The large chunks of endosperm are ground again in the next set of rollers, which the husk bypasses to prevent further damage. Larger malt mills may have three sets of crushing rollers and are capable of effectively handling larger variations in grain size.

Although two-roll milling equipment designed primarily for processing animal feed can be used successfully in the brewery, equipment engineered specifically for brewing is able to produce much more reliable and brew-friendly results. Brewers who transition to mills with multiple sets of rollers generally report marked increases in the ability to recover extract

Fig. 12.1: Two-roll mill in a small brewery. Above the mill is the malt hoppery. A flex auger removes milled malt.

and as a result, may be able to substantially lower the malt quantities used in their beer recipes.

"After we installed the four-roller mill, we were able to reduce our malt charge per brew from 1600 pounds down to 1450," said John Bryant at Starr Hill brewery in Crozet, Virginia. Although the increase in quality may be the main reason to upgrade a mill, for an operation that makes multiple brews per day, the savings in raw material can also help justify the purchase of better milling equipment.

Alec Mull recalls when Founders Brewing Co. in Grand Rapids, Michigan used an older and very rudimentary "brewpub type" mill. "We couldn't get the rollers parallel and that caused us to have five to ten percent "old maids" (unmilled, uncracked kernels)." After a few years the decision was made to upgrade the equipment. Grist preparation at Founders needs to be rapid as they "live mill" directly into the mash tun without an intervening grist case. As part of this speed requirement, they selected a Buhler 4-roller mill. Although the mill cost more than $50,000, it quickly paid for itself in wort extract gains. "When we looked at it in Promash we found that our yield had gone from the high 60s and low 70s to the mid-80s. We were able to take about 10 percent of the grist bill right off the top." The gains quickly created a positive return on investment, and perhaps more importantly, increased quality. "The brewers love it; it is fast; we can mill eight to nine thousands pounds per hour, and there is very little adjustment needed."

Wet Milling

When considering the goal of proper milling is minimal damage to the husk material while still crushing the endosperm, it is beneficial if the husk is pliable and less susceptible to shattering. This can be achieved with "malt conditioning" the practice of adding a small quantity of water to the malt immediately prior to milling. To wet the grain, it is passed through a short length of screw conveyor equipped with spray nozzles. Pre-wetting the grain not only reduces damage to the husk, but also leads to better lauter bed volume and porosity, which in turn allows for a larger amount of grist to be processed in the lauter tun. Malt conditioning has the added bonus of reducing grain dust.

The amount of water used is quite small; about a quart for every hundred pounds of malt. Many brewers would cringe at the thought of bringing water into the milling area, as there are many possible ways that something could go awry when water interacts with rotating equipment, grain, and dust. There are few things in the brewery that are as microbiologically active as grain dust, and it is hard to imagine a more ideal media for beer spoilage organisms than the pasty mud that results when water mixes with that dust. Commercial systems have been designed with this in mind, however, and with frequent cleaning and robust controls, most professional installations can be kept free of any contaminants.

Another type of malt conditioning is "wet milling," which uses much larger qualities of water in the grinding operation. The first wet milling installations included a large tank where warm water was added to the grist and it was subsequently steeped for up to 30 minutes. During the holding time, the water was recirculated from bottom to top to achieve consistency in the batch. Because the malt takes in a lot of water by the time milling starts, the endosperm is practically squeezed out of the husk by a single set of closely spaced rollers. In these systems, the slurry of crushed malt and water arrive at the Malt Conversion Vessel (MCV) at or below the classic protein rest temperature of 122°F (50°C) and are ready to begin the lengthy process of the classic multi-step continental mashing regime.

Steep-Conditioned Wet Milling

Steep-conditioned wet milling blends and refines the two previously mentioned techniques. Roughly 60 seconds prior to arriving at the crushing rollers, the malt enters a steeping chamber where it is exposed briefly to hot water. The husk material absorbs water rapidly, and by the time the malt reaches the rollers, the hulls are pliable enough to resist substantial damage. Additional water is added to the crushing and pumping chambers and the mash arrives at the MCV at conversion temperature. As elements of both milling and mashing are combined in this operation, precise control of water quantity and temperature are absolutely critical for consistent brewing. Because the start of milling also marks the start of the mash conversion period, steep-conditioned wet mills are designed to be large, so that they can rapidly process an entire batch in less than 20 minutes.

Like wet-milling, a benefit to fully hydrated milling is that no dust is created in the process. For any brewer that has spent an afternoon painstakingly seeking out and cleaning every corner in a milling room, that seemingly small factor is a welcome development. In addition, wet milling means the possibility of spark generation or an explosion in the dust-filled milling chamber is nearly nonexistent. These relatively large and complicated machines are equipped with extensive Clean-In-Place (CIP) capabilities. Although most standard milling equipment is constructed out of mild steel, wet mills require more expensive, corrosion resistant alloys.

Many brewers have reported that wet mills, when compared to the same mash profile processed through a dry milling system, have increased the fermentability of wort. For example, Matt Brynildson, brewmaster at Firestone Walker, was initially surprised to see beers that had been finishing at 2.8 Plato decrease to 1.5. (SG 1.011 and 1.006). For a beer that started at 11.5 Plato the respective apparent degree of attenuation increased from 76 to 87 percent! He was able to use this attribute to successfully "dry out" a few beers that they were brewing. "We saw higher enzymatic activity and that drove up the total fermentability. After a few brews we got to where we needed to be by adjusting mash times and temperatures. We feel like we gained more control over our mashing."

Fully-hydrated milling of large quantities of specialty malts and grains is not without its challenges. Because gums and other viscous substances are easily liberated at the high water temperatures, careful attention must be paid to the order and distribution of the grist charge, or the brewer can end

Fig. 12.2: GEA Millstar^{TM} Steep-conditioned wet mill.

up with a hot sticky mess. Larry Sidor, longtime brewmaster at Deschutes Brewery, recalls that when he was involved in the wet milling installation there were a few hiccups that arose when making standards like Black Butte Porter and Obsidian Stout. "I made clay when I put all that high friability dark malt into that machine." Larry vividly recalls the need to frequently disassemble and clean out all of the mash transfer piping. This problem was ultimately solved by distributing the dark malts throughout the grist charge, a technique Larry refers to as "the parfait method."

When the brewhouse equipment includes a mash filter, the grist can be much more finely divided because the husks are not needed to create and maintain a porous filter bed. These grists are nearly unrecognizable to a brewer familiar only with mash/lauter tun operations; they have the consistency of fine cornmeal. Although hammer mills have been the preferred method of preparing grist for mash filters for many years, more recent developments have introduced devices that process the grain into smaller particles using water based, high shear systems. These systems include the rotary disc wet grinding Meura "Hydromill" and Ziemann's wet pump hammer mill, "Dispax."

Grist Analysis

Dry roller mills are used by the vast majority of breweries. For these operations, the quality of the grind can be assessed by sieve analysis— using stacked sieves equipped with progressively smaller screens. By separating a representative grist sample into size based fractions, it is possible to determine the relative weight contribution of each portion. As with most quality tests, a sense of perspective and experience is needed to interpret the results effectively. These tests give insight into the most desired attributes of post-mill malt—consistency and repeatability.

Sieve analysis requires a representative sample of grist. Many mills are equipped with a sampling trier capable of capturing product as it passes the rollers. When using these devices it is important not to fill the sample chambers, as overfilling may lead to an unrepresentative sample. If pulling a sample from the mill is not possible, one can also be pulled from the grist case or conveyor. It is preferable to get the sample from the mill as the effect of any minor process adjustments can immediately be determined. Because shaking methods can vary, commercial equipment such as the Ro-Tap® is available to standardize the procedure.

It is important to codify the duration and vigor of the technique used if conducting the procedure manually.

Grist assortments are commonly separated into four categories: husks, coarse grits, fine grits, and flour (or powder). Although values for "ideal" size assortments are discussed in many texts, most breweries find these values of limited use beyond the initial equipment startup phase. Ultimately, the brewer is looking to mill the grain to make a mash that performs well in the lauter tun, and to achieve the highest quantity of good quality extract in an acceptable time, which in itself, is subjective. What is "acceptable" in one brewery might not work in another, and how each brewery defines wort quality may also vary.

Most breweries use a wide variety of raw materials that are subject to variations in the supply chain. Given the amount of uncontrollable variables, optimizing milling can become a Sisyphean task. Woe to the brewer who strays too close to the minimum mill gap setting for a single lot of poorly modified malt, resulting in a "lauter from hell" story told over many beers. For a brewer using tens of thousands of pounds of malt in every mash, achieving an additional 1 percent yield can add up to significant savings, quickly. A smaller brewery may decide to just add an additional pound of malt to get clearer wort and save 20 minutes in lautering time for the brew.

Regardless of the system used, brewers have a few universal guidelines. If malt quality is poor, efficient milling becomes even more critical. Opening the roller gap will result in lower extract, but that extract will run off easier. Grain dust in the mill room is a problem; aside from being potentially explosive, that extract will not make it into the wort, and dust is tiresome to clean up. For lauter tun operation, grits and husks are desired, flour or powder are not. Powder can point to excessive destruction of the husk material which is particularly high in the polyphenolic compounds that express as tannic harshness in the finished beer.

Milling is the final malt processing step before mashing, and its optimization requires meticulous attention to fine details. The choice of equipment is only the first of many variables. A brewer must also schedule inspections, paying special attention to verification of calibration set points. Over time, mill rollers can and will wear, resulting in inconsistent grist assortment. Maintenance should also include inspection and possible replacement of drive belts and lubrication of the bearing surfaces that the rollers turn on.

The first step for a brewer interested in optimizing their milling operation is to understand where they are starting. Unfortunately it is not possible to separate out the effects of milling from that of mashing and lautering. Because these distinct operations are defined and limited by each other the system must be adjusted together. The situation becomes even more complex when the effect of malt variability is added in.

System efficiency is an excellent indicator of developing problems. If you are not getting the expected quantity of extract out of the mashing process this should trigger some questions. The root cause could be as simple as a leaking valve (a slow trickle of wort down the drain has big impacts) or may result from milling practice or even the malt itself. Regardless, regular review of the calculated ratio between actual and theoretical performance shows a lot. Putting the calculations into a spreadsheet also has the added benefit of providing a regular opportunity to review the malt certificate of analysis.

If you have ruled out wasted wort and malt quality changes then it is probably time to look at the mill. Grist analysis is best done with sieves but simply retaining a sealed sample to visually compare current to standard also has considerable merit. Natural variations in malt make it impossible to develop a single assortment for all situations. Malt made from poor quality barley doesn't mash or lauter well and as a result may need to be less finely ground for it to run through the system in a reasonable time. This will in turn reduce the milling/brewing efficiency.

A brewer who has a good relationship with their malt supplier should get some indication of potential problems in advance and can expect and adjust for these factors instead of needing to react on the fly.

Commercially Available Malts in the US as of 2014

The following list contains a large selection of commercially available malts, by supplier. A small group of specialty producers (known professionally as the Craft Maltsters) are making a wide range of malts using varied grains and production techniques. Because of the rapidly changing, highly varied nature of these small run malts, they are not included in this listing.

This listing includes producer name, malt name, and color where provided.

Malt Company	Malt Name	Average Color (SRM)
Bairds	Pilsen	1.5
Bairds	Maris Otter	3.0
Bairds	Pale Ale	3.0
Bairds	Vienna	3.4
Bairds	Munich	5.0
Bairds	Light Carastan	15.0
Bairds	Carastan (30-40°L)	34.0
Bairds	Dark Crystal	75.0
Bairds	Extra Dark Crystal	135.0
Bairds	Chocolate Malt	475.0

Bairds	Black Malt	550.0
Bairds	Roasted Barley	550.0
Best	Pale Wheat	N/A
Best	Heidelberg Malt	1.4
Best	Chit Malt	1.4
Best	Pilsen	1.8
Best	Acidulated Malt	2.2
Best	Wheat Malt	2.3
Best	Caramel Malt Pils	2.4
Best	Pale Ale Malt	2.7
Best	Smoked Malt	2.8
Best	Vienna	3.7
Best	Munich Light	6.1
Best	Wheat Malt Dark	7.3
Best	Munich Dark	10.0
Best	Red X	12.0
Best	Caramel Malt Light	12.0
Best	Aromatic Malt	19.5
Best	Melanoidin Malt	25.0
Best	Caramel Malt 1	35.0
Best	Caramel Malt Extra Dark (Crystal Malt)	71.0
Best	Special X	133.0
Best	Black Malt	435.0
Briess	Pilsen Malt	1.2
Briess	Carapils® Malt	1.3
Briess	2-Row Carapils® Malt	1.5
Briess	2-Row Brewer's Malt	1.8
Briess	6-Row Brewer's Malt	1.8
Briess	Red Wheat	2.3
Briess	White Wheat	2.5
Briess	Pale Ale Malt	3.5
Briess	Goldpils® Vienna	3.5
Briess	Vienna Malt	3.5
Briess	Rye Malt	3.7
Briess	Smoked Malt	5.0

Briess	Ashburne® Mild Malt	5.3
Briess	Bonlander® Munich Malt	10.0
Briess	Munich Malt 10 L	10.0
Briess	Caramel Malt 10	10.0
Briess	2-Row Caramel Malt 10L	10.0
Briess	Aromatic (Munich) Malt	20.0
Briess	Munich Malt 20 L	20.0
Briess	Caramel Malt 20	20.0
Briess	2-Row Caramel Malt 20L	20.0
Briess	Caramel Vienne Malt 20 L	20.0
Briess	Victory® Malt	28.0
Briess	2-Row Caramel Malt 30L	30.0
Briess	Caramel Malt 40	40.0
Briess	2-Row Caramel Malt 40L	40.0
Briess	Caracrystal® Wheat	45.0
Briess	Special Roast Malt	50.0
Briess	Carabrown® Wheat	55.0
Briess	Caramel Malt 60	60.0
Briess	2-Row Caramel Malt 60L	60.0
Briess	Caramel Munich Malt 60L	60.0
Briess	Caramel Malt 80	80.0
Briess	2-Row Caramel Malt 80L	80.0
Briess	Caramel Malt 90L	90.0
Briess	Caramel Malt 120	120.0
Briess	2-Row Caramel Malt 120L	120.0
Briess	Extra Special Malt	130.0
Briess	Roasted Barley	300.0
Briess	Chocolate Malt	350.0
Briess	2-Row Chocolate Malt	350.0
Briess	Dark Chocolate Malt	420.0
Briess	Black Malt	500.0
Briess	2-Row Black Malt	500.0
Briess	Black Barley	500.0
Briess	Blackprinz® Malt	500.0
Briess	Midnight Wheat	550.0
Canada Malting	Superior Pilsen	1.5

Canada Malting	Distillers Malt	1.8
Canada Malting	Canadian 2-Row	1.9
Canada Malting	Canadian 6-Row	1.9
Canada Malting	Rye	2.5
Canada Malting	Superior Pale Ale	3.1
Canada Malting	White Wheat	3.5
Canada Malting	Munich	8.0
Cargill	Ida Pils™	1.6
Cargill	Euro Pils	1.6
Cargill	German Pils	1.6
Cargill	Schreier Six-Row Pale	1.8
Cargill	Schreier Two-Row Pale	2.0
Cargill	Two-Row Pale	2.0
Cargill	White Wheat	2.9
Cargill	Special Pale	3.5
Cargill	Munich	9.5
Cargill	Caramel 10	11.5
Cargill	Caramel 20	20.0
Cargill	Caramel 30	30.0
Cargill	Caramel 40	40.0
Cargill	Caramel 60	60.0
Cargill	Two-Row Caramel 60	60.0
Cargill	Caramel 80	78.0
Castle	Chateau Pilsen 2RS	1.6
Castle	Chateau Whisky Light®	1.6
Castle	Chateau Whisky	1.6
Castle	Chateau Whisky Light Nature	1.6
Castle	Chateau Whisky Nature	1.6
Castle	Chateau Pilsen 6RW	1.8
Castle	Chateau Pilsen 2RW	1.8
Castle	Chateau Pilsen Nature®	1.8
Castle	Chateau Diastatic	2.0
Castle	Chateau Wheat Blanc	2.1
Castle	Chateau Cara Wheat Nature	2.1
Castle	Chateau Wheat Blanc Nature	2.1

Castle	Chateau Peated (Smoked)	2.1
Castle	Chateau Peated Nature	2.2
Castle	Chateau Chit Barley Malt Flakes	2.4
Castle	Chateau Chit Wheat Malt Flakes	2.4
Castle	Chateau Chit Barley Malt Flakes Nature	2.4
Castle	Chateau Chit Wheat Malt Flakes Nature	2.4
Castle	Chateau Cara Clair	2.5
Castle	Chateau Spelt	2.5
Castle	Chateau Spelt Nature	2.5
Castle	Chateau Oat	2.6
Castle	Chateau Vienna Nature	2.6
Castle	Chateau Vienna	2.7
Castle	Chateau Rye	3.1
Castle	Chateau Rye Nature	3.1
Castle	Chateau Smoked	3.6
Castle	Chateau Smoked Nature	3.6
Castle	Chateau Pale Ale	3.8
Castle	Chateau Pale Ale Nature	3.8
Castle	Chateau Acid	4.0
Castle	Chateau Acid Nature	4.0
Castle	Chateau Buckwheat	4.2
Castle	Chateau Buckwheat Nature	4.2
Castle	Chateau Wheat Munich Light	6.1
Castle	Chateau Munich Light Nature	6.2
Castle	Chateau Munich Light®	6.2
Castle	Chateau Cara Blond®	8.1
Castle	Chateau Wheat Munich 25	9.8
Castle	Chateau Munich	9.8
Castle	Chateau Munich Nature	9.8
Castle	Chateau Melano Light	15.6
Castle	Chateau Abbey Nature®	17.4
Castle	Chateau Abbey®	18.8
Castle	Chateau Cara Ruby®	19.3
Castle	Chateau Biscuit®	19.3

Castle	Chateau Biscuit Nature	19.3
Castle	Chateau Cara Ruby Nature®	19.3
Castle	Chateau Melano	31.0
Castle	Chateau Arome	38.0
Castle	Chateau Cara Gold®	46.0
Castle	Chateau Cara Blond Nature	46.0
Castle	Chateau Cara Gold Nature®	46.0
Castle	Chateau Crystal®	57.0
Castle	Chateau Crystal Nature	57.0
Castle	Chateau Cafe Light®	94.0
Castle	Chateau Special B Nature	109.0
Castle	Chateau Special B®	113.0
Castle	Chateau Cafe	177.0
Castle	Chateau Black of Black	188.0
Castle	Chateau Chocolat	338.0
Castle	Chateau Chocolate Nature	338.0
Castle	Chateau Roasted Barley	432.0
Castle	Chateau Black	497.0
Castle	Chateau Black Nature	497.0
Castle	Chateau Black	507.0
Crisp	Clarity	N/A
Crisp	Process Malt	N/A
Crisp	Naked Oat Malt	1.6
Crisp	Finest Maris Otter	1.7
Crisp	Extra Pale Malt	1.7
Crisp	Glen Eagle's Maris Otter	1.7
Crisp	Europils Malt	1.7
Crisp	Dextrin Malt	1.8
Crisp	Wheat Malt	2.0
Crisp	Clear Choice Malt™	2.3
Crisp	Best Ale Malt	3.0
Crisp	Pale Ale	3.3
Crisp	Extra Pale Maris Otter	3.5
Crisp	Vienna Malt	3.5
Crisp	Light Munich Malt	5.0
Crisp	Cara Gold	6.5

Crisp	Torrified Wheat	7.6
Crisp	Rye Malt	8.0
Crisp	Caramalt	12.5
Crisp	Cara Malt 15	12.5
Crisp	Munich Malt	17.5
Crisp	Dark Munich Malt	20.0
Crisp	Amber Malt	29.0
Crisp	Crystal Light 45	45.0
Crisp	Brown Malt	53.0
Crisp	Crystal 60	60.0
Crisp	Light Crystal	65.0
Crisp	Crystal Dark 77	75.0
Crisp	Medium Crystal	103.0
Crisp	Crystal Extra Dark 120	120.0
Crisp	Dark Crystal	173.0
Crisp	Pale Chocolate	225.0
Crisp	Chocolate Malt	380.0
Crisp	Black Malt	510.0
Crisp	Roasted Barley	510.0
Dingemans	Pilsen	1.6
Dingemans	Pale Wheat	1.6
Dingemans	Organic Pilsen	1.6
Dingemans	Pale Ale	3.3
Dingemans	Munich	5.5
Dingemans	Cara 8	7.5
Dingemans	Roasted Wheat (Tarwe Mout Roost 27)	12.0
Dingemans	Aromatic (Amber 50)	19.0
Dingemans	Biscuit (Mout Roost 50)	23.0
Dingemans	Cara 20	23.0
Dingemans	Cara 45	47.0
Dingemans	Aroma 150	75.0
Dingemans	Special B	148.0
Dingemans	Chocolate (Mout Roost 900)	340.0
Dingemans	De-Bittered Black Malt (Mout Roost 1400)	550.0

Dingemans	De-Husked Roasted Barley	600.0
Fawcett	Lager Malt	1.4
Fawcett	Wheat Malt	1.8
Fawcett	Torrefied Wheat	1.8
Fawcett	Flaked Barley	1.8
Fawcett	Oat Malt	2.3
Fawcett	Maris Otter	2.5
Fawcett	Halcyon	2.5
Fawcett	Pipkin Pale Ale Malt	2.5
Fawcett	Peated Malt	2.5
Fawcett	Golden Promise	2.7
Fawcett	Optic	2.7
Fawcett	Pearl	2.7
Fawcett	Rye Malt	2.8
Fawcett	CaraMalt	14.8
Fawcett	Pale Crystal	30.0
Fawcett	Amber	36.0
Fawcett	Crystal Wheat	54.0
Fawcett	Crystal Malt I	65.0
Fawcett	Crystal Malt II	65.0
Fawcett	Brown Malt	75.0
Fawcett	Crystal Rye	75.0
Fawcett	Dark Crystal Malt I	87.0
Fawcett	Dark Crystal Malt II	150.0
Fawcett	Pale Chocolate Malt	263.0
Fawcett	Roasted Wheat	380.0
Fawcett	Chocolate Malt	500.0
Fawcett	Roasted Barley	600.0
Fawcett	Black Malt	650.0
Gambrinus	Organic Two-Row Pale	1.8
Gambrinus	Organic Pilsen	2.1
Gambrinus	Organic Wheat	2.3
Gambrinus	ESB Pale	3.5
Gambrinus	Vienna Malt	4.0
Gambrinus	Munich 10L	10.0

Gambrinus	Honey Malt	17.5
Gambrinus	Munich 30L	33.0
Great Western	Premium 2-Row Malt	2.0
Great Western	Organic Pilsner	2.0
Great Western	Northwest Pale Ale Malt	2.8
Great Western	Vienna Malt	3.5
Great Western	Wheat Malt	3.8
Great Western	Munich Malt	9.0
Great Western	Organic Munich	10.0
Great Western	Crystal 15	15.0
Great Western	Crystal 30	30.0
Great Western	Crystal 40	40.0
Great Western	Crystal 60	60.0
Great Western	Organic Caramel 60	60.0
Great Western	Crystal 75	75.0
Great Western	Crystal 120	120.0
Great Western	Crystal 150	150.0
MaltEurop	Pilsen	N/A
MaltEurop	Special Kilned	N/A
Malting Company of Ireland	Irish Distillers Malt	1.5
Malting Company of Ireland	Irish Lager Malt	1.8
Malting Company of Ireland	Irish Sout Malt	1.8
Malting Company of Ireland	Irish Ale Malt	2.8
Meussdoerffer	Pilsen	1.7
Meussdoerffer	Vienna	2.5
Meussdoerffer	Munich	5.5
Patagonia Malt	Extra Pale Malt	1.6
Patagonia Malt	Pilsen Malt	2.0
Patagonia Malt	C15	17.0
Patagonia Malt	Caramel 25L	29.0
Patagonia Malt	C35	37.0
Patagonia Malt	C45	45.0
Patagonia Malt	C 55L	57.0
Patagonia Malt	C70	72.0
Patagonia Malt	C90	90.0

Patagonia Malt	C 110L	112.0
Patagonia Malt	Brown 115L	115.0
Patagonia Malt	Especial Malt 140L	139.0
Patagonia Malt	Caramel 170L	168.0
Patagonia Malt	Caramel 190L	193.0
Patagonia Malt	Coffee 230L	230.0
Patagonia Malt	Perla Negra (Black Pearl)	340.0
Patagonia Malt	Chocolate	350.0
Patagonia Malt	Barley 350L	350.0
Patagonia Malt	Black Pearl 415L	410.0
Patagonia Malt	Dark Chocolate	445.0
Patagonia Malt	Barley 450L	445.0
Patagonia Malt	Black Pearl 490L	490.0
Patagonia Malt	Black Malt	530.0
Patagonia Malt	Barley 530L	530.0
Pauls	Pale Ale	3.0
Pauls	Mild Ale (Dextrin Malt)	4.0
Pauls	Caramalt	12.5
Pauls	Amber Malt	20.0
Pauls	Light Crystal	43.0
Pauls	Medium Crystal	60.0
Pauls	Dark Crystal	78.0
Pauls	Extra Dark Crystal	135.0
Pauls	Chocolate Malt	453.0
Pauls	Black Malt	548.0
Pauls	Roasted Barley	640.0
Rahr	Premium Pilsner	1.8
Rahr	Old World Pilsner	1.8
Rahr	Standard 2-Row	1.9
Rahr	Standard 6-Row	2.3
Rahr	Unmalted Wheat	2.8
Rahr	High DP Distillers Malt	2.8
Rahr	Red Wheat	3.3
Rahr	White Wheat	3.3
Rahr	Pale Ale	3.5

Schill Malz	Wheat Malt (White Malt)	1.5
Schill Malz	Pilsner Malt (Pale Malt or Lager Malt)	1.5
Schill Malz	Munich Light Malt	3.1
Schill Malz	Vienna Malt (Amber Malt)	3.5
Schill Malz	Cologne Malt (Kolsch Malt)	4.4
Schill Malz	Munich Dark	6.1
Simpsons	Aromatic Barley	N/A
Simpsons	Caramalt Light	N/A
Simpsons	Pinhead Oats	1.5
Simpsons	Oat Flakes	1.5
Simpsons	Finest Lager Malt	1.7
Simpsons	Pilsner Lager Malt	1.7
Simpsons	Peated Malt	1.7
Simpsons	Low Colour Maris Otter	1.7
Simpsons	Extra Pale Ale Malt	1.7
Simpsons	Distilling Malt	1.7
Simpsons	Wheat Malt	2.1
Simpsons	Maris Otter	2.5
Simpsons	Golden Promise	2.5
Simpsons	Best Pale Ale Malt	2.5
Simpsons	Vienna Malt	3.4
Simpsons	Golden Naked Oats	6.2
Simpsons	Munich Malt	8.1
Simpsons	Caramalt	12.5
Simpsons	Imperial Malt	17.5
Simpsons	Amber Malt	20.0
Simpsons	Premium English Caramalt	23.0
Simpsons	Aromatic Malt	23.0
Simpsons	Crystal Light	40.0
Simpsons	Dark Aromatic Malt	42.0
Simpsons	Crystal Medium	68.0
Simpsons	Crystal Dark	101.0
Simpsons	Simpsons DRC	113.0
Simpsons	Crystal Rye	117.0
Simpsons	Coffee (Brown) Malt	151.0

Simpsons	Crystal Extra Dark	179.0
Simpsons	Chocolate Malt	338.0
Simpsons	Roasted Barley	488.0
Simpsons	Black Malt	497.0
Warminster	Smoked Malt NEW	N/A
Warminster	Maris Otter	3.0
Warminster	Maris Otter	3.0
Warminster	Organic Pale Ale	3.0
Warminster	Pale Ale	3.0
Warminster	Rye Malt	3.0
Warminster	Amber	20.0
Warminster	Crystal 100	40.0
Warminster	Brown	43.0
Warminster	Crystal 200	78.0
Warminster	Crystal 400	133.0
Warminster	Chocolate Malt	500.0
Warminster	Roasted Barley	615.0
Weyermann	Organic Pilsner	N/A
Weyermann	Organic Wheat	N/A
Weyermann	Organic Munich® I	N/A
Weyermann	Organic Munich® II	N/A
Weyermann	Organic Carahell®	N/A
Weyermann	Organic Caramunich®II	N/A
Weyermann	Organic Carafa® II	N/A
Weyermann	Organic Vienna	N/A
Weyermann	Extra Pale Premium Pilsner Malt	1.3
Weyermann	Pilsner	1.9
Weyermann	Barke® Pilsner Malt	1.9
Weyermann	Bohemian Floor Malted Pilsner	2.0
Weyermann	Bohemian Pilsner	2.1
Weyermann	Pale Wheat	2.1
Weyermann	Bohemian Floor Malted Wheat	2.1
Weyermann	Carafoam®	2.2
Weyermann	Acidulated Malt	2.3
Weyermann	Spelt Malt	2.5

Weyermann	Oak Smoked Wheat	2.5
Weyermann	Smoked Malt	2.9
Weyermann	Pale Ale	3.0
Weyermann	Pale Ale Malt	3.0
Weyermann	Rye	3.2
Weyermann	Barke® Vienna Malt	3.4
Weyermann	Vienna	3.4
Weyermann	Light Munich	6.1
Weyermann	Bohemian Floor Malted Dark	6.6
Weyermann	Dark Wheat	7.2
Weyermann	Barke® Munich Malt	7.9
Weyermann	Dark Munich	9.0
Weyermann	Carahell®	10.0
Weyermann	Carabelge®	12.8
Weyermann	Abbey Malt®	17.5
Weyermann	Carared®	19.5
Weyermann	Caraamber®	27.0
Weyermann	Melanoidin	27.0
Weyermann	Caramunich® I	35.0
Weyermann	Caramunich® II	46.0
Weyermann	Carawheat®	48.0
Weyermann	Caramunich® III	57.0
Weyermann	Carabohemian®	74.0
Weyermann	CaraAroma®	151.0
Weyermann	Chocolate Rye	244.0
Weyermann	Roasted Rye, unmalted	244.0
Weyermann	Carafa® I	338.0
Weyermann	DeHusk Carafa® I	338.0
Weyermann	Chocolate Wheat	395.0
Weyermann	Roasted Wheat, unmalted	420.0
Weyermann	Carafa® II	432.0
Weyermann	DeHusk Carafa® II	432.0
Weyermann	Carafa® III	526.0
Weyermann	DeHusk Carafa® III	526.0
Weyermann	Sinamar®	3120.0

Worldwide and North American Malthouse Capacities

North American Malthouse Capacity by Location

Company	Location	Metric Tons/Year
Briess Malt & Ingredients Co.	Chilton, WI	15,000
Briess Malt & Ingredients Co.	Waterloo, WI	30,000
Busch Agricultural Resources	Idaho Falls, ID	320,000
Busch Agricultural Resources	Moorhead, MN	92,000
Cargill Malt	Biggar, SK	220,000
Cargill Malt	Sheboygan, WI	30,000
Cargill Malt	Spiritwood, ND	400,000
Gambrinus Malting Co.	Armstrong, BC	6,200
GrainCorp (Canada Malting)	Calgary, AB	250,000
GrainCorp (Canada Malting)	Montreal, QC	75,000
GrainCorp (Canada Malting)	Thunder Bay, ON	120,000
GrainCorp (Great Western)	Pocatello, ID	92,000
GrainCorp (Great Western)	Vancouver, WA	120,000
InteGrow Malt	Idaho Falls, ID	100,000
MaltEurop	Great Falls, MT	200,000
MaltEurop	Milwaukee, WI	220,000
MaltEurop	Winnipeg, MB	90,000
MaltEurop	Winona, MN	115,000
MillerCoors	Golden, CO	230,000
Rahr Malting Co.	Alix, AB	140,000
Rahr Malting Co.	Shakopee, MN	370,000

World Largest Commercial Malting Compaies

FIRST KEY — CONSULTANTS TO THE BREWING AND MALTING INDUSTRIES

June 2014

R	WPS	Company	Total	Breakdown (location: capacity)
1	9.8%	Malteries Soufflet	2,148	France: 116, 91, 53, 81, 72, 241; Czech Republic: 59, 53, 44, 100, 108; Germany: 86, 61, 54; Romania: 106; Poland: 115; B*: 26; Russia: 112; Ukraine: 160; Kaz.: 85; Serb.: 75; Brazil: 105
2	9.7%	Malteurop Groupe	2,138	France: 241, 82, 55, 36; Germany: 100; Spain: 155; Port.: 60; Pol.: 42; China: 55, 25; Ukraine: 112, 58; Russia: 110; Poland: 160; USA: 115; Australia: 200; Canada: 90; Austr.: 75; N.Z.: 42
3	9.7%	Cargill Malt	2,126	USA: 440; Canada: 30, 105; Russia: 110; Spain: 100; Argentina: 330; Australia: 200, 80, 96, 12, 45, 8, 110
4	6.1%	GrainCorp Malt	1,333	USA: 120, 93; Canada: 250; 80; United Kingdom: 53, 45, 83, 31; Germany: 80, 60, 15; Australia: 103, 23, 46, 86
5	5.0%	Boortmalt Gr. Axéréal	1,093	France: 165; Belgium: 330; Hung.: 29, 75; United Kingdom: 175, 51, 60, 59; Ireland: 94; Cro.: 55
6	3.9%	Russky Solod Gr. Avangard	864	Russia: 128, 128, 140; Germany: 130, 70, 55, 85
7	3.6%	Supertime	800	China: 320, 300, 50, 60, 50, 20
8	3.5%	Cofco Malt	760	China: 360, 320, 80
9	2.5%	Rahr Malting	540	USA: 400; Canada: 140
10	1.8%	Beidahuang Longkien	400	China: 100, 200, 100
11	1.6%	Viking Malt	360	Sweden: 220; Finland: 75; Lith.: 65
12	1.6%	Ireks	348	Germany: 65, 20, 18, 23, 48, 100; Austria: 74
13	1.4%	Dalian Xinge	300	China: 300
14	1.3%	Simpsons Malt	286	United Kingdom: 236, 50

E.U.-27
Europe non EU
North America
South America
Australia/N.Z.
China - India

R = Rank
WPS = World Production Share
Quantities in 000 tons

| World Production 2013 (est.) | 22,000,000 mt |
| World Capacity 2013 (est.) | 26,700,000 mt |

Top 3 Malting Companies	29.1%**
Top 5 Malting Companies	40.2%**
Top 10 Malting Companies	55.5%**
Top 20 Malting Companies	68.3%**

**of world est. production

*Bulgaria

© *Daniel Huvet, Managing Director, Agribusiness Group.*

R	WPS	Company					
15	1.3%	Bar Malt	80	India 100	100		280
16	1.2%	Sihai Chengde	China 270				270
17	1.2%	Chunlei	50	60	China 50	45 50	255
18	1.1%	Crisp Malting Gr.	United Kingdom 115	40	35	30 30	250
19	1.1%	Hanse-Malz	60	Germany 75	110		245
20	1.1%	Holland Malt	Netherlands 130	105			235
20	1.1%	Agromalte Agraria	Brazil 235				235
22	1.0%	Xin Jinwei	85	China 55	70 20		230
23	1.0%	The Malt Company	India 150	30 45			225
24	0.9%	Global Malt	Germany 110	Pol. 85			195
25	0.8%	Muntons	United Kingdom 95	80			175
26	0.8%	Muogao	China 100	70			170
27	0.7%	Fuglsang	Denmark 110	47			157
28	0.5%	Maltexco	Chile 31	34 35			100

Craft Maltsters Listing

North American Craft Maltsters

Company	Location	Annual Capacity, in Tons
Academy Malt Co.	Indianapolis, IN	80
Blacklands Malt	Austin, TX	107
Blue Ox Malthouse	Belfast, ME	-
Christensen Farms Malting Co.	McMinnville, OR	68
Colorado Malting Co.	Alamosa, CO	600
Corsair Artisan Distillery	Nashville, TN	100
Doehnel Floor Malting	Victoria, BC	11
Eckert Malting & Brewing Co.	Chico, CA	20
Farm Boy Farms	Pittsboro, NC	125
Farmhouse Malt NYC	Newark Valley, NY	50
Grouse Malting & Roasting Co.	Wellington, CO	100
Hillrock Estate Distillery	Ancram, NY	100
Malterie Frontenac Inc.	Quebec, Canada	825
Mammoth Malt	Thawville, IL	30
Michigan Malt	Shepherd, MI	50
New York Craft Malt	Batavia, NY	156
Niagara Malt	Cambria, NY	50
Our Mutual Friend Malt & Brew	Denver, CO	1
Pilot Malt House	Jenison, MI	30

Rebel Malting Co.	Reno, NV	40
Riverbend Malt House	Asheville, NC	200
Rogue Ales Farmstead Malthouse	Newport, OR	15
Sprague Farm and Brew Works	Venango, PA	10
Valley Malt	Hadley, MA	300
Western Feedstock Technologies, Inc.	Bozeman, MT	20

Introduction To Home Malting

By George de Piro

Reprinted with permission from Zymurgy.

Malting your own grain is labor intensive, time consuming, and infinitely more fun and educational than reading a book (or magazine article!) about it. An in-depth knowledge of malt is key to formulating outstanding beers. To truly understand the malting process one must get one's hands dirty and actually do it.

Like commercial malting, home malting can be separated into three basic steps: steeping, germination, and kilning.

Steeping is performed to bring the relatively dry grain to a moisture content of about 45 percent so that germination can commence. During this phase the grain will be alternately submerged in water and then drained and allowed to rest. The sequence and timing of these phases vary based upon the character of the barley and the preferences of the maltster. Monitoring the moisture content of the grain is critical during this phase.

During germination, growth of the tiny barley plant begins inside the seed and roots sprout and grow on the exterior. Physical and chemical changes take place during this growth that make the kernel suitable for use in brewing. While moisture content is still important during this phase, the goal will be to achieve a certain degree of growth. This is assessed by checking to see how much progress the acrospire or barley shoot has made in growing from the root end of the kernel toward the tip. In low-modification malts, it will cover only one-half to two-thirds of the distance; in well-modified malts, three-quarters or more will be covered.

Kilning dries and toasts the grain, halting growth and imparting many of the flavors we associate with malt. In most cases, drying occurs first at lower temperatures (100–120°F or 37.7–48.8°C) and toasting proceeds only after moisture content has been reduced to about 10 percent. To a large extent, the temperature of toasting determines the final character of the malt.

The equipment you need to accomplish each of these steps and produce your own malt is in large part dependent on the amount of malt you wish to produce. You can make a pound or two using small plastic containers and other common kitchen items. For larger amounts, up to 15lb (6.8kg) or so, your malt can be made using stuff that most all-grain brewers already possess. Here is a basic equipment list:

Scale: A scale with the ability to accurately measure mass up to 200g in 0.1 g increments is useful for moisture determinations. A scale with larger capacity can be used to measure grain and malt.

Steep tank: This can be a 5-gallon, foodgrade plastic bucket with holes drilled into the bottom placed into another 5-gallon bucket without holes drilled in the bottom. The old "Zap Pap" lauter tun works perfectly!

Malting floor: Aluminum roasting pans work well, as would any shallow, flat pan or plastic container. If you have a particularly clean basement floor, you could try just spreading the malt on it. Most home maltsters will opt for a container of some sort.

Household space heater: Useful for low-temperature kilning. For small batches, food dehydrators can be used.

Household fan: A fan is useful for drying malt at low temperatures prior to kilning.

Kiln: A kitchen oven can be used successfully, but temperature control is likely to be laborious and imprecise. Still it is the best most of us can hope for. There are reports of people using clothes dryers, but I have no experience with them (other than their obvious use).

Thermometer: An accurate thermometer with a temperature range of at least 45–212°F (7–100°C) is very useful. A higher range will enable you to make more accurate temperature measurements when making crystal and roasted malts.

Commercial and home malting are theoretically similar, but there are some important differences. While each lot of barley must be treated differently regardless of size, small-scale maltings can germinate much faster than larger batches. This may be due to the intensive aeration that is possible when malting small amounts of grain. Malting schedules must therefore be looked upon as guidelines rather than gospel. It is important to use your senses of taste, smell, touch, and sight to determine when to move on to the next phase. The one objective analytical tool that can help you monitor the progress of your malt is moisture content. Before we move on to discuss the three phases of malting, let's discuss this important procedure.

Moisture Content Determinations

The moisture content, also referred to as the degree of steeping, can be determined in two ways. The first is to take a sample of grain from the batch, weigh it, dry it and then weigh it again. This technique can be used at any time and at any phase of the malting process. Short of burning the kernels during drying, it is fairly foolproof. We'll call this the "drying method."

The second method that can be used is to entrap a small sample of grain in a perforated container (called a Bernreuther apparatus) that is included in every step of the process. By weighing the grains before processing begins and knowing their initial moisture content, you can directly determine moisture content by weighing them again at any point in the process. This technique depends on two things. First, you have to maintain exactly the same population of kernels in the container throughout the process. Second, the grains in this sample must receive exactly the same treatment as the rest of the batch so that they are representative of the whole batch. We'll call this the "direct method."

In both systems for assessing moisture content, we will be working with the same equation:

Equation 1:
(weight of moist grain − weight of dry grain) / weight of moist grain x
100 = % moisture content

Using the drying method, a sample is accurately weighed and then placed in an oven on a baking sheet or similar device in a thin layer and heated at 212–220°F (100–104°C) for three hours (Note: the grain should not become brown or burnt during this procedure—if so, your oven may be too hot). After the drying is complete, weigh the grain again and use the values you have obtained in the equation.

Using the direct method, you would first determine the moisture content of your barley using the drying method. The sample of grains used for this purpose would be discarded. Next you would put some barley in your Bernreuther apparatus (the perforated container), remembering that during steeping the grains will swell to occupy nearly 50 percent more space than when dry. Once you have selected the sample, weigh it and then return it to the apparatus. You will now know the moisture content of your barley and the weight of your initial sample. In order to do calculations using Equation 1 during the malting process, you will need to calculate the dry weight of your sample using equation 2.

Equation 2:
sample weight x (1 - moisture content as a decimal) = dry weight of sample

Once you begin the malting process, you will be able to remove the Bernreuther apparatus from the batch, open it, weigh the grains and then return them to the apparatus and the batch in process. The weight you determine each time will give you the "weight of moist grain" needed for use in equation 1. You will use the value for "dry weight of sample" from equation 2 for the "weight of dry grain" value in equation 1.

The primary value of the direct method is that it allows very rapid assessment of current moisture levels during malting whereas the drying method requires a three-hour delay. Also, when small batches are being produced the drying method may result in the loss of a significant amount of grain by the end of processing.

The Malting Phases

Now that you are familiar with the main quantitative measure used to aid in malting, we are ready to discuss the individual phases of the operation.

Steeping is performed to bring the relatively dry grain to a moisture content of about 45 percent so that germination can commence. Water uptake will be influenced by several factors including: steeping time, temperature of steep water, kernel size, barley variety, and character.

Steeping consists of two stages: wet steeps and air rests. During the wet steeps the grain is covered with clean, cool water. During the air rests, the water is drained from the grain to allow for respiration of oxygen and removal of carbon dioxide.

The length and number of steeps and rests can vary widely based upon the character of the barley and the maltster's preferences. Indeed, most maltsters conduct a series of pilot maltings on small samples before beginning to malt a production-size batch. This helps them to determine the best steep/rest schedule and germination conditions.

Now, here's my basic procedure for steeping using a Zap-Pap double-bucket style mashtun. The grain is placed in the bucket that has holes drilled in the bottom. This bucket is then placed into the "unholy" bucket. The grain is covered with cool water (50–55°F) and rinsed with a continuous overflow of water for about 15 minutes to remove debris. After the grain is clean enough for your tastes, it is left covered with cool water to steep. After an hour, the interior bucket is removed from the other and set down. The oxygen-depleted steep water is dumped out and the wet grain is poured back and forth between the buckets several times to ensure thorough aeration. It may then be covered with fresh, cool water again.

This aeration should be performed every hour for the first few hours of the first wet steep. The moisture content of the grain can be assayed at the end of the steep and may be as high as 30 percent. After the grain has been steeped, the water is drained off, the grain is turned, and it is then allowed to remain in the steep vessel without water for the first air rest.

During the air rest the grain continues to absorb the moisture adhering to it and germination begins. The respiring grain will generate a fair amount of heat and carbon dioxide and may become dry to the touch. Frequent turning and rinsing with cool water will keep the grain aerated and moist.

Be sure to smell, feel, and taste the grain during this process. The grain should not smell or taste sour or rancid at any time. It should taste clean and grainy. As germination begins it will take on an odor similar to cucumbers or unripe apples. This is your sign that everything is going well.

Near the end of steeping, the grain will show the first signs of germination, namely chitting. Chitting is when you see a small white spot or bump at the broad end of the barley kernel. This whitish structure is the rootlet beginning to emerge.

Once your grain achieves the target moisture level it is time to move on to germination.

Germination in traditional maltings occurred on a malting floor. At home, you are not likely to want to spread malt all over your house to allow it to germinate. Not only would this be of questionable sanitation, but the people you cohabitate with may be justifiably annoyed, and your dog will find the malt delicious.

Shallow aluminum roasting pans or plastic bins (available at fine super-markets and hardware stores everywhere) are ideal for germinating small quantities of grain. Transfer the moist grain to the malting pans in layers about 2" (5cm) deep and watch the fun unfold.

During germination, the rootlets which began to emerge in the steep tanks grow rapidly. To keep them from tangling into an inseparable clump, the malt must be gently mixed and turned at least twice a day. Also, the grain must be misted with cool water frequently to maintain the desired moisture content. Finally, the temperature of the grain should be maintained in the range from 55–65°F (12.8–18.3°C).

The temperature during germination has a big effect on the quality of the malt. Those practicing floor malting tend to keep the grain cooler than their counterparts using modern methods. Although germination proceeds more rapidly at warmer temperatures, those who practice floor malting believe that cooler temperatures yield a malt of higher quality. Commercial maltsters must therefore strike a balance: they want to use the warmest temperature that will produce quality malt in order to speed production of the malt.

Speed and economics should not be a concern for the home maltster. Low temperatures help to ensure even moisture distribution within the seed and even modification. Steeps should be kept around 50–55°F while the germination should also be kept as close to 55°F as possible. Frequent turning of the malt, keeping the germinating malt in a thin layer, misting with cool water, and keeping the malt in a cool room will all work to realize this goal. The germinating grain can be as much as 10°F warmer than the air temperature, even when spread in a thin layer, so it is important to keep track of the temperature of the grain rather than the air.

The embryonic plant, or acrospire, grows using some of the energy stored in the starch of the endosperm to fuel its development. In wet grain, the acrospire can be visualized beneath the husk on the dorsal side of the kernel. The amount of acrospire growth is related to the degree to which the biochemical changes occur in the grain. These changes are referred to as modification.

The degree of acrospire growth is related to the degree of modification. The longer the acrospire, the more modified the malt. Maltsters usually halt germination when the acrospire length is between 75–100 percent of the length of the malt.

To determine average acrospire length, you will want to select a small handful of kernels and determine where the acrospire is in each one. In damp germinating malt, you can usually see the acrospire quite clearly through the barley husk. Ideally, what you will find is that most of the acrospires are about the same length, but this is not always the case. At home you may find that your grains have widely differing acrospire lengths. If you simply allow the slowest pieces to achieve acrospire growth of 75 percent of total length, the faster pieces will be overgrown. If the acrospire is allowed to grow too much, it will consume a lot of the starchy endosperm to fuel its growth, substantially reducing the extract available to the brewer. As a result, it is best to examine a cross section and put a halt to germination when the average hits about 75 percent.

If you have a significant portion of the grains with short acrospire growth, you may need to use a protein rest or even a decoction mash to maximize extract from the grain during mashing. Since almost all of today's commercial malts are well-modified and need no protein rests, producing your own under-modified malt is a way to more accurately replicate brewing techniques and beers of old.

The amount of time it takes for the grain to become adequately modified is highly variable with a range from a few days to as much as a week. The grain should be inspected frequently to determine when it is time to halt germination—a step that is achieved by drying the grain.

Kilning conditions are determined by the type of malt that is being made and the restrictions of your home kiln. Diastatic malts such as Pilsner and pale are dried to a moisture content of about 10 percent at a relatively low temperature before being kilned off at 150–185°F. It is important that the malt is fairly dry before the temperature is raised to preserve the enzymes.

The most reliable way to do this at home is to simply dry the malt at room temperature until it is at a moisture content of about 10 percent.

A space heater and household fan can be used to heat a room to 80–90°F and blow warm air across the green malt. Depending on the relative humidity, the malt will dry to 10 percent moisture in a day or so. Use the moisture determination assay to track moisture content.

Once the malt is at 10 percent moisture, it is ready to kiln at higher temperatures. The desired color of the malt determines the kilning temperature used. Most pilsner malts are kilned off at no higher than 185°F (85°C) for four to eight hours. Deeper color and toasted flavor are developed by kilning at higher temperatures for longer periods of time.

Kilning the grain at too high a temperature for too long will reduce (or eliminate) its diastatic capacity. If the grain is still moist when kilned, the damage to diastatic enzymes will be even greater. The malt must be turned during drying and kilning to ensure even temperatures.

Of course the green malt does not need to be kilned at all. The lightest-colored malt achievable is simply dried at warm room temperature. This very pale malt is called wind malt or sun malt depending on your climate. While this practice was followed by early homebrewers, it has a few shortcomings. First, you will probably not be able to dry the malt enough to ensure biological stability during storage. Thus unkilned malts should be used within a few weeks of production. Second, there is a flavor impact as much of the cucumber-like green malt character that is normally kilned off will remain in the wind malt. Whether or not this is desirable is up to the brewer that will use the malt.

Crystal malts are a bit more difficult to make. The wet malt is sealed in a container and heated to saccharification temperature (145–155°F, 63–68°C) until it tastes sweet (one to four hours). This saccharifies the starch in the endosperm in the exact manner that occurs in the mash tun. The sweet grain is then heated to a higher temperature and allowed to dry by increasing ventilation. The higher the temperature, the deeper the color of the crystal malt. The interior of the grain will take on a glassy appearance and be hard to the tooth if all goes well. Thus far, I have not perfected a technique for making crystal malt at home. The crystal malts I produced were a bit withered looking, but tasted wonderful. Fresh crystal malt is something every brewer must experience!

Toasted malts, like Victory malt, and deeply roasted malts like chocolate and black malts are produced by kilning malt at higher temperatures—although care must be taken to ensure that you do not ignite or char the malt.

Munich-type malts are more difficult to produce at home because it is necessary to exercise relatively fine control over the kilning temperatures and moisture content of the malt. The manufacture of Munich-type malts is closely related to that of crystal malts, but they are treated in a manner that preserves much of their diastatic ability.

The procedure for Munich malt outlined in Kunze's Technology Brewing and Malting explains that the green malt is dried to about 25 percent moisture at no more than 104°F (40°C) before being heated to 140–149°F (60–65°C) over a period of nine hours. The malt is then cooled to 122°F (50°C) and allowed to dry to about 12 percent moisture. It is then heated back to temperatures up to 220°F (105°C) to develop the malty-tasting melanoidins and dry the malt to the final moisture content of about 3 percent.

Acrospire Removal

To avoid a bitter or astringent taste in your beer, dry, malted grain should undergo a final separation from the rootlets and acrospires. A quick home method for removing this chaff is to rub the malt around in a kitchen strainer, allowing the finer material to fall out. With larger amounts of malt, it may be quicker to fill an empty pillowcase with malt and bang that around on a hard surface until the chaff comes off, then use the strainer to separate it.

Conclusion

Malting at home is not quite as easy as brewing, but the experience and knowledge you will gain are priceless. Malt is the chief influence on beer color and a critical component of beer's flavor profile. Increasing your knowledge of malt by making it yourself can only make you a better brewer, and if prohibition should ever again rear its ugly head, you'll be that much better prepared!

The author gives many thanks to Roger Briess and Jim Basler at Briess Malt & Ingredients Company for their support of his home-malting endeavors. They provided grain, knowledge, laboratory analyses, and chocolate-covered malt balls that were invaluable.

EXAMPLE STEEPING SCHEDULES

For German 2-row Barley

	Time/Temp	Moisture Content at end of step
Wet Steep	4 hr @ 54° F (12° C)	32%
Couch	20 hr @ 63° F (17° C)	34%
Wet Steep	4 hr @ 54° F (12° C)	38%
Couch	20 hr @ 70° F (21° C)	40%
Wet Steep	2 hr @ 59° F (15° C)	44%

Home Malting of Harrington 2-row Barley

	Time/Temp	Moisture Content at end of step
Wet Steep	11 hr @ ~50° F (10° C)	35%
Couch	3 hr @ ~70° F (21° C)	—
Wet Steep	6 hr @ ~50° F (10° C)	38%
Couch	5 hr @ ~70° F (21° C)	—
Wet Steep	11 hr @ ~50° F (10° C)	42.5%
Couch	3 hr @ ~70° F (21° C)	—
Wet Steep	4 hr @ ~50° F (10° C)	43.5%

In all schedules, wet steeps must include aeration every one to two hours. Couch phase must include CO2 removal every two to three hours.

Bibliography

American Malting Barley Association (AMBA). July 2014. *No Genetically Modified (GM) Varieties Approved for Commercial Production in North America*. Milwaukee, WI. http://ambainc .org/content/58/gm-statement.

Anderson, P.M., E.A. Oelke, and S.R. Simmons. 1985. *Growth and Development Guide for Spring Wheat. University of Minnesota Agricultural Extension Folder AG-FO-2547*.

Baker, Julian L. 1905. *The Brewing Industry*. London: Methuen & Co.

Bamforth, Charles W. 2002. *Standards of Brewing: A Practical Approach to Consistency and Excellence*. Boulder, CO: Brewers Publications.

Bamforth, Charles W. 2006. *Scientific Principles of Malting and Brewing*. St. Paul, MN: American Society of Brewing Chemists.

Barnard, Alfred. 1977. *Bass & Co., Limited: As Described in Noted Breweries of Great Britain & Ireland*. Burton upon Trent: Bass Museum. Sir Joseph Causton and Sons.

Baverstock, James, and J. H. Baverstock. 1824. *Treatises on Brewing*. London: Printed for G. & W.B. Whittaker.

Beaven, E. S. 1947. *Barley, Fifty Years of Observation and Experiment*. Foreword by Viscount Bledisloe. London: Duckworth.

Beschreibende Sortenliste. 2011. Bundessortenamt. Hannover: Dt. Landwirtsch.-Verlag.

Bickerdyke, John. 1886. *The Curiosities of Ale & Beer: An Entertaining History*. London: Field & Tuer.

Blenkinsop, P. 1991. "The Manufacture, Characteristics and Uses of Speciality Malts", *MBAA Technical Quarterly*, Vol. 28(4), 145-149. St. Paul, MN: MBAA.

Briggs, D. E. *Malts and Malting*. 1998. 1st ed. London: Blackie Academic and Professional.

Briggs, D.E, J.S. Hough, R. Stevens, and T.W. Young. 1981. *Malting and Brewing Science*. London: Chapman and Hall.

Clark,Christine. 1978. *The British Malting Industry Since 1830*. London, U.K. Hambledon Press.

Clark, George & Son Ltd. 1936. *Brewing: A Book of Reference*. Volumes 1, 2, 3,4,5,6. London.

Clerck, Jean de. 1957. *A Textbook of Brewing Vol. 1. Vol. 1*. [S.l.]. London: Chapman & Hall.

Clerck, Jean de. 1958. *A Textbook of Brewing Vol. 2. Vol. 2*. [S.l.]. London: Chapman & Hall.

Colby, C., 2013. "German Wheat Beer III, Mashing and the Ferulic Acid Rest": http://beerandwinejournal.com/german-wheat-beer-iii/.

Combrune, Michael. 1758. *An Essay on Brewing With a View of Establishing the Principles of the Art*. London: Printed for R. and J. Dodsley in Pall-Mall.

Cook, A. H. 1962. *Barley and malt biology, biochemistry, technology*. New York: Academic Press.

Coppinger, Joseph. 1815. *The American Practical Brewer and Tanner*. New York: Van Winkle and Wiley.

Covzin, John, 2003. *Radical Glasgow: A Skeletal Sketch of Glasgow's Radical Traditions*. Glasgow: Voline Press. http://www.radicalglasgow .me.uk/strugglepedia/index.php?title=Glasgow,_City_of_Rebellion.

Daniels, Ray. 1996. *Designing Great Beers: The Ultimate Guide to Brewing Classic Beer Styles*. Boulder, CO: Brewers Publications.

Davies, Nigel. 2010. "Perception of Color and Flavor in Malt". *MBAA Technical Quarterly*. Vol. 47. St. Paul, MN: MBAA. Doi:10.1094 /TQ-47-4-0823-01.

Ellis, William. 1737. *The London and Country Brewer*. The 3rd ed. London: Printed for J. and J. Fox.

Fincher, G.B. and Stone, B.A. 1993. "Physiology and Biochemistry of Germination in Barley." *Barley: Chemistry and technology*. eds. A.W. MacGregor and R.S. Bhatty. St. Paul, MN: American Association of Cereal Chemists, Inc. 247– 95.

Ford, William. 1862. *A Practical Treatise on Malting and Brewing*. London, U.K. Published by the Author.

Forster, Brian. 2001. "Mutation Genetics of Salt Tolerance in Barley: An Assessment of Golden Promise and Other Semi-dwarf Mutants". *Euphytica*. 08-2001, Volume 120, Issue 3, Dordrecht, Netherlands: Kluwer, 2001. 317-328.

Foster, T. and B. Hansen, "Is it Crystal or Caramel Malt?" *Brew Your Own*, Nov. 2013.

Fuller, Thomas. 1840. *The History of the Worthies of England*. London, UK: Nuttall and Hodgson.

Gretenhart, K. E. 1997. "Specialty Malts." *MBAA Technical Quarterly* Vol. 34 (2), 102-106. St. Paul, MN: MBAA.

Gruber, Mary Anne. 2001. "The Flavor Contributions of Kilned and Roasted Products to Finished Beer Styles." *MBAA Technical Quarterly*, Vol. 38. St. Paul, MN: MBAA.

Hardwick, William A. 1995. *Handbook of Brewing*. New York: M. Dekker.

Harlan, Harry V. "A Caravan journey through Abyssinia", *National Geographic*, Volume XLVII, No. 6. June 1925.

_____. 1957. *One man's life with barley, the memories and observations of Harry V. Harlan*. New York: Exposition Press.

Harrison, William. 2006. *Description of Elizabethan England, 1577*. Whitefish, MT: Kessinger.

Hayden, Brian, Neil Canuel, and Jennifer Shanse. 2013. "What Was Brewing in the Natufian? An Archaeological Assessment of Brewing Technology in the Epipaleolithic". *Journal of Archaeological Method and Theory*. 20 (1): 102-150.

Hertsgaard, Karen. "Declining Barley Acreage", *MBAA Technical Quarterly*, Vol. 49, No. 1, 2012, pp. 25-27. St. Paul, MN: MBAA.

Hieronymus, Stan. 2010. *Brewing with Wheat: The 'Wit' and 'Weizen' of World Wheat Beer Styles*. Boulder, CO: Brewers Publications.

_____. 2012. *For the Love of Hops: The Practical Guide to Aroma, Bitterness, and the Culture of Hops.* Bolder, CO & Brewers Publications

Hind, H. Lloyd. 1940. *Brewing: Science and Practice.* London: Chapman and Hall.

Hopkins, Reginald Haydn, and Bertel Krause. 1937. *Biochemistry Applied to Malting and Brewing.* London: G. Allen & Unwin Ltd.

Jalowetz, Eduard. 1931. *Pilsner Malz.* Wien: Verl. Institute für Gärungsindustrie.

Johnson, D. Demcey, G. K. Flaskerud, R. D. Taylor, and V. Satyanarayana. 1998. *Economic Impacts of Fusarium Head Blight in Wheat.* Agricultural Economics Report No. 396, Department of Agricultural Economics. Fargo: North Dakota State University.

Jones, B.L., 2005 "Endoproteases of Barley and Malt." *Journal of Cereal Science*, Vol. 42, 139-156.

Katz, Solomon H., Fritz Maytag. 1991. "Brewing an Ancient Beer". *Archaeology.* 44 (4): (July/August 1991), 22-33.

Kawamura, Sin'itiro. "Seventy Years of the Maillard Reaction." *ACS Symposium Series.* (April 29, 1983). American Chemical Society: Washington, DC. doi: 10.1021/bk-1983-0215.ch001. Accessed Oct. 28, 2012.

Kunze, Wolfgang, Hans-Jürgen Manger, and Susan Pratt. 2010. *Technology: Brewing & Malting.* Berlin: VLB.

Lancaster, H. M. 1936. *The Maltster's Materials and Methods.* London: Institute of Brewing.

Leach, R., et al, 2002. "Effects of Barley Protein Content on Barley Endosperm Texture, Processing Condition Requirements, and Malt and Beer Quality", *MBAA Technical Quarterly*, 39(4). 191-202.

Lekkas, C., Hill, A.E., Stewart, G.G., 2014 "Extraction of FAN from Malting Barley During Malting and Mashing", *Journal of the American Society of Brewing Chemists*. 72(1):6-11.

Loftus, W. R. 1876. *The Maltster: A Compendious Treatise on the Art of Malting in All Its Branches*. London: W.R. Loftus.

MacGregor, A.W., Fincher, G.B. 1993. *Barley Chemistry and Technology*, Chapter 3 - Carbohydrates of the Barley Grain, American Association of Cereal Chemists.

McCabe, John T, and Harold M. Broderick. 1999. *The Practical Brewer: A Manual for the Brewing Industry*. Wauwatosa, WI: Master Brewers Association of the Americas.

McGee, Harold. 1992. *The Curious Cook*. London: HarperCollins.

Moffatt, Riley. 1996. *Population History of Western U.S. Cities & Towns*, 1850-1990. Lanham: Scarecrow. 90.

Morrison, W.R., 1993. *Barley Chemistry and Technology*, Chapter 5 - "Barley Lipids." American Association of Cereal Chemists.

Mosher, Randy. 2009. *Tasting Beer: An Insider's Guide to the World's Greatest Drink*. North Adams, MA: Storey.

Ockert, Karl. 2006. *Raw Materials and Brewhouse Operations*. St. Paul, Minn: Master Brewers Association of the Americas.

Omond, George William Thomson. 1883. *The Lord Advocates of Scotland*. Edinburgh: Douglas.

O'Rourke, T. 2002. "Malt Specifications & Brewing Performance." *The Brewer International*. Volume 2, Issue 10.

Palmer, John J., and Colin Kaminski. 2013. *Water: A Comprehensive Guide for Brewers*. Boulder, CO: Brewers Publications.

Pearson, Lynn. 1999. *British Breweries–An Architectural History*. London, U.K.: Hambledon Press.

Piperno DR, E. Weiss, I. Holst, D. Nadel. 2004 "Processing of Wild Cereal Grains in the Upper Paleolithic Revealed by Starch Grain Analysis." *Nature*. 430: 670-673.

Preece, Isaac. 1954. *The Biochemistry of Brewing*. Edinburgh: Oliver & Boyd.

Priest, Fergus Graham, and Graham G. Stewart. 2006. *Handbook of Brewing*. Boca Raton: CRC Press.

Quaritch, Bernard. 1883. *The Corporation of Nottingham, Records of the Borough of Nottingham:1399-1485*. Published under the authority of the Corporation of Nottingham. London.

Riese, J.C., 1997. "Colored Beer as Color and Flavor." *MBAA Technical Quarterly*, Vol. 34(2), 91-95. St. Paul, MN: MBAA.

Scamell, George, and Frederick Colyer. 1880. *Breweries and Maltings; Their Arrangement, Construction, Machinery, and Plant*. London: E. & F.N. Spon.

Scheer, Fred. 1999. "Specialty Malt from the View of the Craft Brewer." *MBAA Technical Quarterly*, Vol. 36(2):215-217. St. Paul, MN: MBAA.

Schwarz, Paul, Scott Heisel & Richard Horsley. 2012. "History of Malting Barley in the United States, 1600–Present" *MBAA Technical Quarterly* vol. 49 (3). St. Paul, MN: MBAA.

Sebree, B.R., 1997. "Biochemistry of Malting", *MBAA Technical Quarterly*, 34(3) 148-151. St. Paul, MN: MBAA.

Serpell, James. 1995. *The Domestic Dog: Its Evolution, Behaviour, and Interactions with People*. Cambridge, U.K.: Cambridge University Press.

Sharpe, Reginald R. (editor). 1899. *"Folios 181–192: Nov 1482. Calendar of letter-books of the city of London: L: Edward IV-Henry VII"*. British History Online. http://www.british-history.ac.uk/report.aspx?compid=33657.

Shewry, P.R. 1993. *Barley Chemistry and Technology*, Chapter 4 - Barley Seed Proteins. American Association of Cereal Chemists.

Simpson, W. J. 2001. "Good Malt – Good Beer?" Proceedings of the 10th Australian Barley Technical Symposium. Canberra, Australia.

Steel, James. 1878. *The Practical Points of Malting and Brewing*. Glasgow, Scotland. Published by the Author.

Strong, Stanley. 1951. *The Romance of Brewing*. London: Review Press for the Crown Cork Co.

Sykes, Walter John, and Arthur L. Ling. 1907. *The Principles and Practice of Brewing (Third Edition)*. London: Charles Griffin & Co.

Thausing, Julius, Anton Schwartz & A.H. Bauer. 1882. *The Theory and Practice of the Preparation of Malt and the Fabrication of Beer*. Philadelphia: H.C. Baird & Co.

Thatcher, Frank. 1898. *Brewing and Malting Practically Considered*. Country Brewers' Gazette Ltd., London, U.K.

Tizard, W. L. 1850. *The Theory and Practice of Brewing Illustrated*. London: Gilbert & Rivington.

Tyron, Thomas. 1690. *A New Art of Brewing Beer, Ale, and Other Sorts of Liquors*. London: Printed for Tho. Salusbury.

Vandecan, S.; Daems, N.; Schouppe, N.; Saison, D.; Delvaux, F.R. (2011). "Formation of Flavour, Color and Reducing Power during the Production Process of Dark Specialty Malts." *Journal of the American Society of Brewing Chemists*. 69 (3), 150-157.

Van Hook, Andrew. 1949. *Sugar, its Production, Technology, and Uses.* New York: Ronald Press Co.

Wahl, Arnold Spencer. 1944. *Wahl Handybook.* Chicago: Wahl Institute, Inc.

Wahl, Robert and Max Henius. 1908. *American Handy Book of the Brewing, Malting, and Auxiliary Trades, Volume Two.* Chicago: Wahl-Henius Institute.

White, Chris, and Jamil Zainasheff. 2010. *Yeast: The Practical Guide to Beer Fermentation.* Boulder, CO: Brewers Publications.

Wigney, George Adolphus. 1823. *A Philosophical Treatise on Malting and Brewing.* Brighton, England: Worthing Press.

Index

carboxylic acid, 104, 107

Carey, Dan, 161–62, 168–69, 175

Cargill Malthouse, 76, 222, 233, 234, xxi. *See also* Schreier Malthouse

caryopsis, 132

cask conditioned mild ales, 121

Castle malthouse, 222–24

catty flavor, 25

Celiac disease, 106–7

cellulose, 95, 101, 103

Centennial barley, 172

cereal. *See also specific cereals*
history of human use of, 27–28
taxonomic classification of, 141
viability of seeds, xvi

cereal flavor, 127

Certificate of Analysis (COA)
Carey on, 168, 185–86
data in, 179–187
description of, 19
dextrinizing power test on, 111
DP listed in, 110
example, **178**
grain bills and, 14, 19
in performance review, 217
purpose of, 177
at receipt of malt, 194
testing laboratories for, 179
using, 185–86

chaff, 194, 201

chain disc conveyers, 198–200

char flavor, 21, 22

character, 50, 118, 240, 242–43

Charles barley, 152, 171

cherry wood, 80, 123, 150

Chevalier, John, 163

Chevallier barley, 163, 169

Chile, 235

chilling, 43

China, 142, 234–35

chit malt, 35, 126

chits, xxiv, 56. *See also* rootlets

chitting, 243–44

chocolate flavor, 109, 123, 128

chocolate malts
basis of, 71
burnt flavor and, 21, 122
char flavor and, 21
color of, 122
description of, 122
dryness and, 21
excessive, 24
home malting of, 247
producing, 79
in robust porters, 21
for stout, 25
"white malt" and, 117

Chocolate Rye, 21, 231

cholesterol, 107

Christensen Farms Malting Co., 237

Chunlei, 235

cider, 36, 163

Cigar City Brewing Co., 20–22

cisterns, 35

clean malt, 21, 30

cleaning
of barley, 52–53, 243
for dust control, 203, 212–13, 216
of grain handling system, 204–5
of malt, 65, 201
solvents used for, 197
of storage areas, 196–97
of wet milling equipment, 212–15

total, 179
unfermentable, 10

falling number viscosity test, 138–39, 152
Farm Boy Farms, 237
Farmhouse Malt NYC, 237
Farrell, Andy, 177
fatty acids, 107–8
Fawcett malthouse, 226
feed, animal, 65, 143–45, 158, 203
"Feed Pro" weighing system, 202
Feekes' growth scale, **130**
Fehling's solution, 110
fermentability, xvi
fermentables, 10, 106, 213
fermentation
    amylase activity and, 111–12
    carbohydrates and, 10, 96
    de-branching enzyme and, 112
    dextrins and, 96
    dust and, 202
    FAN and, 181, 187
    sugars and, 10, 111
ferns, 31
fertilizer, 143–44, 158
ferulic acid, 103, 132
figgy flavor, 21
filtration, 180–81, 187
final gravity (FG), 16
Fine Grind Dry-Basis (FGDB), 182–83
finish, 118, 121, 213
finished malt, 11, 34, 42, 64, 66, 71
finishing, 24
finishing malts, 24, 25
Finland, 234

Firestone Walker Brewing Co., 116, 213
Firlbeck barley, 165
flaking, 81–82
flavor. *See also* taste; *specific flavors*
    adjuncts and, 82
    of amber malt, 121
    of biscuit malt, 121
    of black malts, 70, 122
    of brown malt, 121
    Canadian Brewing and Malting Barley Research Institute studies on UK malts and, 50
    from caramel malts, 69–70, 74, 77
    carbohydrates and, 10, 183
    of chocolate malts, 122
    COA on, 180, 183
    of crystal malts, 120
    debittered dark malts and, 26
    definition of, 178
    descriptors for, 126
    development of, xvi
    from dry hopping, xv
    enzymes and, xvi
    floor malting and, 50
    of green malt, 246
    kilning and, xvi, 51, 62–64
    Maillard reactions and, xxv, 62, 183
    of Maris Otter barley, 170
    modification and, 62, 74
    from oats, 125
    of pale malts, 117
    of peated malts, 124
    of pilsner malts, 117
    quantifying, 26